杨修憬 主编
Chief Editor: Yang Xiujing

中国美术学院建筑遗产保护
国际论坛论文集

Anthology Of The International Architecture Forum:
The Preservation Of Monumental Heritage

中国美术学院出版社
CHINA ACADEMY OF ART PRESS

目录
Contents

1   前言 | 杨修憬
2   Preface | Yang Xiujing

5   水乡又水城——江南水乡南浔古镇之水城再造构想 | 邵健

16   德国的村镇更新 | 鲁道夫·吕克曼
20   Village Renewal in Germany | Rudolf Lückmann

26   建筑遗产的"可利用"保护策略——以广西鼓鸣寨滨水区保护与更新为例 | 沈杰、张蕾

32   通过保护和旅游发展政策复兴托斯卡纳地区的古镇 | 亚历桑德鲁·梅里斯、奥鲁夫托·尔加土图依
35   Regeneration of the Historical Villages of Tuscany, Through Conservation and Tourism Development Strategies | Alessandro Melis, Olufunto Ijatuyi

41   困境与期望——中国传统村落文化保护与活化路径的实践探索 | 孙以栋、郑渊晗

47   以真实性为教条——呼吁以实用的方法保护历史建筑 | 诺伯特·施密茨
51   Authenticity as Dogma – A Plea for the Pragmatic Conservation of Monuments | Norbert M. Schmitz

59   历史古镇的复兴和农业区域的干预措施 | 马利欧·皮萨尼
61   The Restoration of Historic Towns and Interventions in Agricultural Areas | Mario Pisani

67   城市建筑"遗产"再生 | 康胤

74   文明将孕育出新的文明——卢克索地区文化遗产及可持续发展面临的挑战 | 阿哈姆德·拉希德
82   The Civilization will be Born from the Womb of Civilization: Luxor Heritage and the Challenges of Sustainable Development | Ahmed Rashed

95	新型城镇化背景下浙江小城镇人居环境营造之探索	郑军德、徐艺文

101	《修复指南》在建筑遗产保护中的意义及相关研究——以意大利古镇切尔瓦拉·迪·罗马为例	法比欧·匹斯科普

106	"留住一片乡愁"——浙西霞山古镇民居文化及保护对策	陈凌广

111	保存树根以孕育森林——南特案例研究	劳伦特·莱斯考普
119	Shaping the image of a city Nantes a case study	Laurent Lescop

131	历史城镇再生的一种方法——马鞍山采石古镇城市更新	吴晓淇

141	同筑共赢，重构美丽江南——杭州拱墅胜利河古水街项目案	周胤斌

152	"GreenUP - A Smart City" 亦被称为建筑遗产保护理念	贾科莫·皮拉兹·皮拉左里
155	"GreenUP - A Smart City" Aka Preservation to be Conceptualised	Giacomo Piraz Pirazzoli

159	新区新建筑与历史文脉——以杭州双城国际项目为例	项亚量

166	凸显城市身份定义的景观实践教学之道——以杭州工业遗存景观更新设计为例	黄晓菲

174	解析德国威廉皇帝纪念教堂的场所体验——用现象学的方法	程银

前言

杨修憬｜中德学院院长

　　2014年3月，我国国务院公布了《国家新型城镇化规划（2014—2020）》。该规划明确指出，我国现已进入全面建成小康社会的决定性阶段，正处于经济转型升级、加快推进社会主义现代化的重要时期，也处于城镇化深入发展的关键时期。因此有专家分析认为，随着城镇化的发展，与之相关的诸多行业将有望提振。浙江的城镇化率位于全国前列，众多产业也已开始升级转型，相应地对原有城镇的公共服务和生活环境提出了新要求，这势必促使城镇建设的进一步提升。

　　在城镇化建设中，要素的空间聚集、土地的合理运用、生态环境的保护、建筑文化遗产的规划与更新，都离不开科学的研究与规划。近年来我省城市更新和改建的业务量增长相当迅速，而且项目要求的深度和广度也在不断增加，由前些年的外立面改造到目前的大范围城镇拆迁、改建以及旧城保护等。面对大量的改建项目，不少人士还缺少理念和广阔视野，特别是如何处理好保护与开发利用的关系，是设计师面临的主要难题。

　　在许多国家，建筑遗产保护已成为建筑领域的一个重要发展方向，并且其研究的边界不断得到拓展。研究人员往往来自城市规划、建筑设计、景观设计、工业设计、艺术设计、交互设计、考古学、人类文化学、材料学、历史学等不同领域，他们对同一课题进行交叉研究，通过跨学科、跨领域的探索来形成建筑遗产保护的综合方案。这种多学科的研究方式为建筑遗产保护和利用提供了无限的可能性。在世界范围内，德国的城镇化发展经验得到了普遍的肯定，特别在可持续发展的城市群和城市交通、乡村城镇化建设和历史文化遗产保护等方面积累了丰富的经验，值得我们研究与学习。

　　浙江历史悠久，古村落老建筑众多，比如杭州有西湖和大运河两个世界文化遗产保护项目。为全省城镇化建设培养既掌握建筑规划知识，了解建筑遗产保护方法，同时又熟知国际与国内各类文化保护法规的高级专业人才，应该是中德学院办学的一个重要方向。为了借鉴国外在建筑遗产保护领域的成功经验，更好地实现我们的培养目标，由中国美术学院主办、中德学院承办的"建筑遗产保护"国际论坛于2015年12月5日至6日在中国美术学院象山校区召开。论坛旨在拓展我院当代中国本土建筑学的研究视野，实现建筑艺术与人文研究的特色建构。来自5大洲7个国家13所高校的近20位教授、学者和建筑师与会，他们各自展示了在建筑遗产保护领域的实践案例，受到与会听众的高度评价。大家认为，通过分享他们在建筑遗产保护领域的宝贵经验，将有助于我们探索解决我国在城镇化过程中出现的文化遗产保护不力，城乡建设缺少特色等问题的路径。

　　两天的论坛发言，专家们不仅为我们带来了建筑遗产保护领域最新的研究成果，更重要的是通过论坛的召开，为我们今后的合作办学与项目研究搭建了国际化专家平台。以此为起点，我们将发挥我院在艺术人文与艺术设计领域的特色与优势，结合我省实际需要，通过引进国际先进理念，研究以往建设中景观结构与区域自然地理不协调，"建设性"破坏不断蔓延，自然与文化个性被破坏，导致文化传承断流等问题，切实担负起为地方培养高质量人才，将教育主动融入地方、服务地方经济的社会责任。

　　本书收录了与会的外籍教授、学者以及我国高校教师在建筑遗产保护领域的研究成果。相信他们的研究将为我院的建筑遗产保护专业教学与实践提供有价值的样本。

　　在此，要特别感谢安哈尔特应用技术大学副校长、建筑系建筑遗产保护专业主任鲁道夫·吕克曼教授和该校校长助理陆以理女士的大力支持，由于他俩的不懈努力，才使我们有机会完成此书的出版，并为历史留下各地学者的思考痕迹。

2016年11月15日

Preface

Yang Xiujing | Dean of the Chinesisch-Deutsche Kunstacademie

The State Council of the People's Republic of China published New Urbanization Plan (2014—2020) in March, 2014, in which, it was stated that China had entered a decisive stage of completing a moderately prosperous society and was during an important period of economic transformation and upgrading and speeding up socialist modernization and during the critical period of deep development of urbanization. Some experts think that a number of related industries are helpful to boost with the development of urbanization. The urbanization rate in Zhejiang is at the forefront of the country and many industries have been begun to upgrade and transform. Therefore, accordingly, there are new requirements for public services and living environment of original cities and towns, which certainly will prompt further improvement of construction of cities and towns.

In the urbanization construction, elemental spatial aggregation, proper use of land, protection of ecological environment, and planning and update of architectural cultural heritage cannot be separated from scientific research and planning. In recent years, the business volume of updating and rebuilding cities in our cities has increased rapidly and the depth and breadth required by projects have been on the increase from the transformation of façades in previous years to the urban demolition and reconstruction and conversation of old cities within large range at present. Faced with a large number of reconstruction projects, many professionals still lack ideas and broad vision, especially how to deal with the relationship between the protection and the development and utilization, which is the main problem faced by many designers.

In many countries, architectural heritage protection has become an important development direction in the building field, the research range is expanded gradually, and the researchers come from different fields, such as urban planning, architectural design, landscape design, industrial design, artistic design, interactive design, archaeology, cultural anthropology, materials science and history, who conduct cross-over research on a same task to form a comprehensive pan for architectural heritage protection by interdisciplinary exploration. This multi-disciplinary research method provides infinite possibilities to protection and utilization of architectural heritage. The urbanization development experience in Germany has been affirmed generally in the world, especially it has rich experience on the sustainable development of urban agglomeration and urban traffic, rural urbanization construction and the protection of historical and cultural heritage, which is worthy of our research and study.

Zhejiang has a long history and many historic villages and old buildings, and Hangzhou has two world cultural heritage protection projects: West Lake and the Grand Canal, so it is an important direction for Sino-German University to train highly specialized personnel mastering architectural planning knowledge, understanding architectural heritage protection methods and familiar with international and domestic cultural protection laws and regulations for the urbanization in whole province. In order to use foreign successful experience of architectural heritage protection for reference and realize our training objective better, the International Forum of Architectural Heritage Protection sponsored by China Academy of Art and undertaken by Chinesisch-Deutsche Kunstacademie was held in Xiangshan campus of China Academy of Art from December 5, 2015 to December 6, 2015 in order to broaden our universities' research horizon on native Chinese architecture and realize the characteristic construction of architectural art and humanistic research. Near 20 professors, scholars and architects from 13 colleges and universities of countries of 5 continents were present, who showed the practical cases of architectural heritage protection respectively, which were highly praised by the audiences attending the conference. All of them thought that the valuable experience of architectural heritage protection shared by them would be helpful to us to explore the approaches and solve the problems of ineffective cultural heritage protection and lack of features on rural and urban construction during the urbanization process in our country.

In the forum lasting for two days, experts brought newest research results of architectural heritage protection. What's more, they built an international expert platform for our further cooperation in running the school and project research. With this as a beginning, we will give full play to the characteristics and advantages of our university in art, humanity and artistic design, combined with actual requirements in our province and introducing international advanced ideas, to research the problem of cutoff of cultural inheritance caused by discordance between landscape structure and regional physical geography, constant extension of "constructive" damage, destroy of natural and cultural individuality and others in the past construction so as to effectively shoulder the social responsibility of training high-quality talents for local place and actively blending the education in local place to serve local economy.

The research results of foreign professors and scholars and college teachers in our country on architectural heritage protection have been collected in this book. We believe their research will provide valuable samples for the teaching and practice of architectural heritage protection major of our university.

Hereon, we would like to thank the strong support of Professor Rudolf Lückmann from the Architectural Heritage Protection Major of Department of Architecture of Anhalt University of Applied Sciences and Ms. Lu Yili, whose unremitting efforts contribute to the publication of this book and leave the thinking trials of all scholars for the history.

November 15, 2016

邵健
Shao Jian

中国美术学院建筑艺术学院副院长。研究方向为景园规划与设计，长期专注于传统聚落与园林营造的学研，并运用于城市开放空间的设计实践，近年来从事规划、建筑与景观的一体化设计。所主持课程曾获教育部"国家级教学成果奖二等奖"。代表作品有陈之佛艺术馆、中国美术学院附中综合楼等，多项作品曾获省勘协一、二等奖及全国美展优秀奖。

1992年　毕业于中国美术学院环境艺术系并留校任教至今。
2014年　"龙泉大师园徐朝兴陶艺工坊"获浙江省第十三届美展艺术设计金奖。

Professor, Vice-Director of School of Architecture, China Academy of Art
Graduated in Department of Environment Art, China Academy of Art in 1992, Shao Jian has since stayed at school as a teacher. His research direction is the planning and design of garden and he has always focused on the reseach and study of traditional settlements and garden building, which he then applied to the design of urban open space.He engages in the integrated design of planning, architecture and landscapein recent years. His courses won the second award of National Teaching Achievements by the Ministry of Education. His representative works, for example Chen Zhifo Museum of Art, the Comprehensive Building of China Academy of Art High School and so on, have won the first and second awards issued by Zhejiang Provincial Exploration & Design Association and the Excellence Award of National Art Exhibition Award; His work Xu Chaoxing Pottery Workshop in Longquan Master Gardenwon the Golden Award of Art Design in the 13th Zhejiang Province Art Exhibition in 2014.

水乡又水城
——江南水乡南浔古镇之水城再造构想

邵健

【摘要】近三十年经济高速发展使"江南水乡"特征渐趋消失,显著的表征是河道锐减,水生态危机,居民对水的依赖与情感变得迟钝。本文从这样一个普遍性却又异常艰难的现实问题单刀直入,尝试破解水乡之困局。通过长达三年的南浔田园调查与文献研究,确定以南浔水系的修复为切入点,进而探讨古镇全面复兴的相关策略,期望具有长远意义的江南水乡的真实回归。

[Abstract] The rapid economic development in recent 30 years causes the features of watery towns in southern China to disappear gradually. The obvious representations are sharp decrease of rivers, water ecological crisis, bluntness of residents' dependency and emotion on water. We try to crack the dilemma of watery regions in this article from this universal and very difficult practical problem directly. We take the restoration of Nanxun river system as an entry point after three years of survey and document research on Nanxun countryside to discuss the strategies related to the overall revival of ancient towns for the real return of watery towns in southern China with long-term significance.

一、水乡之殇

提起江南水乡,一幅水网纵横,河湖交错,田园村舍,粉墙黛瓦的淡彩水墨画便呈现在眼前,"小桥、流水、人家",朴素而雅致。江南水乡,历朝历代所指地域各有不同,但江南似乎是一种象征,象征着山清水秀的一方土地,文人雅士聚集,象征着繁盛富庶的鱼米之乡。而本文的江南,相对狭义,指的就是太湖流域,植根于"水网"环境中的吴越文化体系,具有温婉秀美的自然景观以及富足安乐的生活特征。特别是隋唐开挖的京杭大运河成为中国南北交通大动脉后,农业水利和交通得到很大的发展,呈现"苏湖熟,天下足"的经济繁荣。时至今日,这里仍是中国经济文化最发达的地区之一。

然而,在社会经济高速发展和城市化的进程中,大量的河道被道路代替,许多河流出现断流甚至消亡的现象,河网水系呈现主干化态势,河道功能衰退。另一方面,工业废水和城市生活污水大量排放,农业污染不断上升,水生态系统自身的净化能力丧失,水资源污染问题突显。此外,古镇内基础配套设施缺乏,年轻人逐步趋向城市,留下的主要是老年人,水乡习俗逐渐淡出,水乡文化岌岌可危。如此诸多问题,使太湖流域的古镇走向群体性衰弱,曾经惬意的水乡生活与我们渐行渐远。

太湖流域小镇众多(图1),发展的同时,均面临城市基础设施有待改善,产业提升与转型等问题。如何实现传统小镇的全面复兴是一道难题,目前仅靠观光式的旅游只会使小镇趋于景观化,非复兴之根本。水乡,终归以水为魂,水乡生活的继承与发展是根本,否则只能徒有虚名。因而,以水系修复与整治为复兴开端似乎是一条必然之路。本文

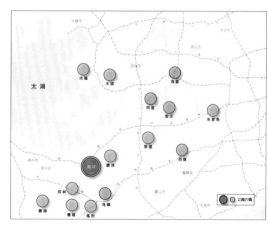

图1

以此为基点，进而推导江南古镇南浔的水城再造的宏观构想。

二、南浔之辉煌

南浔位于中国长三角城市群的中心腹地太湖南岸，是大运河江南段的重要市镇（图2、图3），南浔镇第二批进入"中国历史文化名镇（村）"。

南浔建镇于南宋淳佑季年（公元1252年），取南林、浔溪两名之首字命名，南浔一称至今已约760年。明末，南浔已是"烟火万家"的"江浙之雄镇"。清道光、咸丰年间，南浔辑里湖丝获首届世博会金奖，声誉国际市场，"蚕事吾湖独盛，一郡之中，尤以南浔为甲"，丝市商贸繁荣，产生了中国近代最大的丝商群体。据史料记载，南浔丝商群体当时的总资产达白银六千万两至八千万两之间；光绪年间，南浔，一个江南小镇的库银竟然与全国库银的三分之一等，甲富天下，名不虚传。

南浔儒商"经世致用，回报社会"，以近代传奇人物张静江为代表，倾家族之财力资助孙中山辛亥革命，"当时出资最勇而多者张静江也"，他是国民党的四大元老之一，蒋介石尊称其为"革命导师"。南浔崇文收藏之风颇盛，名扬内外。如刘承干的"嘉业藏书楼"，藏书60万卷，其中《永乐大典》珍贵孤本42册，《四库全书》（翁覃溪手纂）原稿150册，尤为珍贵。嘉业藏书楼（图4）不但收藏丰富，而且还刻版印书，为传承中华文化所作的贡献，是当时国内所有私家藏书楼无可比拟的。另有张钧衡的"六宜阁"，藏书10万卷；蒋汝藻的"密韵楼"，藏有宋元善本两千余部。1919年国学大师王国维曾应邀在密韵楼研校群书达4年之久，可见藏书之精。再看书画收藏，"虚斋"主人庞莱臣，被誉为"全世界最负盛名的中国书画收藏大家"，其收藏的历代名画不仅数量达几千件，且至精至美，品味极高。庞莱臣在文化收藏史的意义其实不亚于齐白石和黄宾虹的意义。还有西泠印社创始人之一的张石铭，中国泰斗级书画鉴定大家张葱玉等，其家族藏品令人叹为观止。

吴兴园林（图5、图6）自南宋开始渐盛，在中国园林史上有一席之地。至近代园林，童寯先生予以高度评价："以一镇之地，而拥有五园，且皆为巨构，实江南之仅见……吴门之外，此当首推矣。"还有中西合璧的江南大宅，与园林一体，其

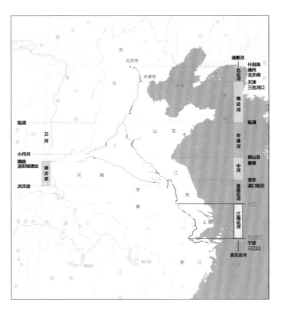

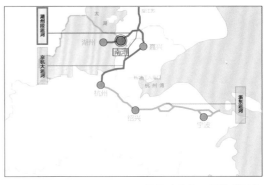

图2（上），图3（下）

图4（左上）
图5（右上）
图6（左下）
图7（右下）

规模宏大与营造精美，是近代中国建筑史上的一个地方缩影，也是海派文化历史传统的见证，为后人留下了难能可贵的文化遗产。（图7、图8）

限于篇幅，数例重要人事以佐南浔"民国第一镇"实不为过。浔商志存高远、积极开放、兼容并蓄的精神深刻影响着该地区的人文基因，如何继承并发扬光大，是城市复兴策略应予积极回应的。

三、水生活

儿时记忆中，居民日常饮水都取之河水。天色微亮，起床的第一件事就是下河埠打水，河水经过一夜的沉淀，清澈见底。两岸居民提着木桶、铁桶陆陆续续到河埠，撇开水面浮物，快速沉下，提起，这样往返于灶头与河埠数次，大致一缸，是一家当天的用量。之后，才到河边淘米、烧粥，灶头里架上桑木后，利用空档，搭上毛巾到河边洗漱。这时河埠开始热闹起来，沿河望去，十足的生活味。到太阳快出来的时候，基本是洗洗刷刷的家务活了。邻里间的话也多了起来，隔岸寒暄。摇橹船划来，坐满了人，是各村"出街"赶集的；也有一人小渔船划来，半船的水，装着鲜笼活跳的各种河鱼，都是清早捕的，岸边想买的随时喊住，论斤两或个数；也有临水楼阁里探出头来的，从窗扇外直接用绳子放下竹篮，装着钱，渔夫便放上鱼，一声吆

图8（左），图9（右）

喝，提回楼里。如此种种，可谓"水市千家聚，商渔自结邻，长廊连箬屋，斥堠据通津"。

不仅夜间的溪水星月，水中人家，无不闪光透亮，而且白昼更加富丽。阳光逐波而生辉，照彻一切，而生七色。从帆船竹筏，菱藕鱼虾，到木石津梁以至两岸树林、行人、街坊，没有不是被水波荡漾着，被波光亮晶晶地闪耀并似从三棱镜中，折射出虹彩来的。这种水乡生活的灵动景象，深深印在每个水乡人的骨子里，难怪报告文学大家徐迟连用六十六个"水晶晶的……"来赞叹家乡南浔的灵秀风韵。

时至今日，生活节奏快了，但居民与水为邻的情节依然延续，午茶晚餐，纳凉或是家务，依然喜欢沿河摆摊；客栈，也总是临水的生意好；即便深宅大院，依然喜欢装点戴望舒"雨巷"的情调（图9）。即使行色匆匆的路人，依然会在途中不自觉地找个临水石凳，或桥栏上小歇一会，陌生人照样搭话，吴侬软语里夹杂着隐约传来的书场评弹，颇具风韵——也许这就是水乡那种自然生长着的魅力吧。

南浔古镇复兴期盼这种"笃定"的水乡气质得以保持，延续。

四、水系演变与城市变迁

南浔水源自西天目山的东西苕溪，由于地势向东趋向低平，水由西东流，分二支流经南浔，北部頔塘西承东苕溪之水入境，南部简五塘承西苕溪德清方向水流，二股水流汇合于十字港，北泄太湖，东下黄浦江。"十字形"水系构成了南浔古镇的主要骨架，形成一个典型的水市。其中，横穿全境的頔塘是太湖流域开凿最早的运河之一，也是南浔镇的交通与经济命脉。

明初，南浔城墙拆毁，建造了东南西北四个水栅，从此，南浔镇傍水而成的特点更趋明显。由城壕、百间楼河等水系围合形成的"申"字水系又成为南浔的重要特征。至道光年间，南浔镇区"自东栅至西栅三里之遥，距运河（在北）而至南栅五里"，略小于府城，水系结构清晰，因水成市特征明显。

1954年，因頔塘不能满足日益发展的水运要求，绕过市镇，开挖了长湖申新航道，从此运河畅流。1973年填平北市河加阔宝善街，水系演变成"丁字形"，实则影响镇区河港泂流，对水利颇有影响。此后，南浔水系的末端河流不断减少，河网密度下降，水系结构趋于主干化和简单化，结构表征指标表现为逐年下降。对比

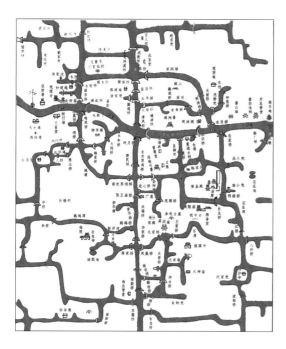

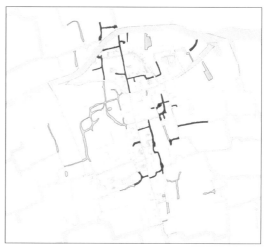

图10（左上）
图11（右上）
图12（右下）

现在、1981年、1908年三张水系示意图（图10、图11），从中可以一窥水系式微的历程。

南浔由已"十字形"河流形成市镇。"水"是古镇生态系统的血脉所在，以水为脉络延展布局，从而衍生出与水生活相关的各种文化品性、风俗习惯等，沿河街市成为聚集人气、突显古镇生命力的场所。

五、水系修复之必要性

频塘故道开凿于晋代，是太湖流域最早的运河之一，距今已有1700余年历史。这是一项规模庞大的古代水利工程。2014年6月，中国大运河项目列入"世界遗产名录"。大运河（湖州段）列入遗产范围是南浔的频塘故道，遗产点为南浔历史文化街区，总面积2.18平方千米（图12）。因各朝各代的整修与拓浚，今天，频塘故道依然发挥着水利设施和航运通道的功能。这样特殊的古河道，如何进行保护，继续发挥其在社会发展中的作用，有着更高的要求，而古镇水系的修复将使这一目标得以

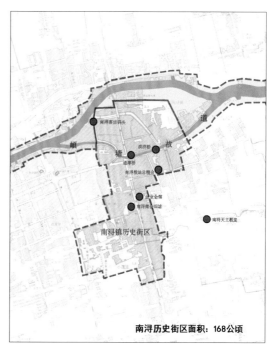

实现。

与太湖流域其他古镇一样，南浔现状问题凸显，尤其是车行交通无法有效到达，无法满足生活与业态需求，因此，河道重新作为交通手段变得分外迫切。古镇保护范围东界至宜园遗址东侧起，西界至永安街西，南界自嘉业堂藏书楼及小莲庄起，北界至百间楼，保护面积约1.68平方千米，其中重点保护区面积0.88平方公里，现存河道约5500米，其中水域面积约0.09平方千米。通过考证原有的河道消失的原因、位置，发现大部分河道主体仍在，如宝善街、凤凰河，当初填埋的方式直接，北市河就在现状道路之下，其他填埋河道情况类似，南浔街巷格局与建筑现状是有利于水系的原真性修复的。

无论从大运河遗产保护的要求，还是城市复兴的内在诉求来看，水系修复迫在眉睫。

六、真实性修复规划

在国家文物局《中国大运河（201303）》申报世界遗产文本中，提到了大运河的遗产完整、环境良好和通航顺畅前提下的保护和利用。中国城市规划设计研究院的《南浔古镇及周边地区城市设计研究规划（2013）》及湖州市城市规划设计研究院的《南浔历史文化保护区保护规划（2004）》等上位规划，也都提到历史水系的修复的必要性，可见真实性修复目标是一致的。

首先是十字水系的北市河修复，关系水城的结构的完整性。北市河在上世纪70年代被填埋，现为宽16至20米的宝善街，据挖掘考证（图13），河道位置清晰，宽度基本在6至8米之间，东岸距房屋7至10米，西岸宽度3至5米，相对窄些，留存的老房子位置表明，东侧街道部分为临水建筑，西岸是临水街道。两岸石砌驳岸基本完好，部分驳岸有二至三层的痕迹，反映了不同时期百姓对河道的修整乃至侵占的事实，这提醒我们采取更加细致的态度，分析原有河道的连续性、完整性，并对比史料，作出更准确的判断。老人对当时北市河的描述，与实地挖掘考证情况基本一致。这些都为水系的原真性修复提供了依据。

宝善街目前大都是七八十年代的简约板式楼房，少数为清末或民国建筑，传统水乡的风貌并不明显，但却是一个时代的真实状态，呈现出一种独特的现代水乡印记，与南市河、百间楼等历史街区共同彰显出小镇的历史厚度。

南市河分运河水入清风桥，称北市河，北经太平桥、盐店桥、栅桩桥，与西木行港及百间楼港交错。修复后的北市河与南市河贯为一体，长约2.23千米（核心区内），与保存完好的1.8千米頔塘运河，形成一横一纵的十字城市骨架（图14）。

其次是"申"字水系，"十"字的外围一环，北有百间楼港、西木行港，东南是皇御河，西部水

图13

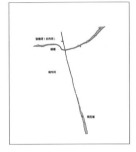 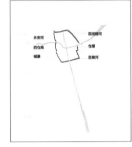

图14（左），图15（右）

系已填埋，即原来的永安港、西仓港、城濠。填埋的永安港目前是9米宽的道路，西侧是近30年建造的民房，规划建议拆除，这有利于城市主道路西移、永安港修复。修复的方法将与北市河相同，找到河岸基址，作为依据，使永安港成为古镇西北区通航的主河道；已经填埋的西仓港位于建设中的垂虹公园西侧，鉴于公园建筑布局，并不影响水系的开挖；最难是城濠，历史上的西南护城河，由于泰安路的主交通及新建的晶街项目阻断了河道的连续，近期修复可能性不大，但作为规划，依然需要控制水系用地，以备长远发展。皇御河北段因宽度较窄无法满足通航，规划勾连了皇御河与仓潭，解决北接頔塘的难题。这样，修复的"申"字河道形成环状水网，相比目前仅仅满足水上游的 "丁"字水系，申字水系是水城的主脉，目标指向破解车行交通带来的小镇宜居困局（图15）。

马家港是东部平行南市河的一条河流，北起頔塘故道，南达望津河，全长0.9千米。便民桥以北的马家港两岸，传统建筑肌理与风貌特征保护较好。而南部马家港，虽然历史资源不足，但仍有传统民居数栋，古桥一座，河埠与百年古树相伴、两岸树木生长茂盛，间杂着蔬果地，有市郊村落生机勃勃的水乡景象，是不同于中心市镇的另一种水乡体验（图16）。马家港与南市河仅200米之隔，两岸建筑密度相对较低，又在保护区外，是高密度古镇突围的主要方向，也是城市土地"增量盘活存量"的重要途径。重要的是，南市河居民的疏导能够实现就近安置，符合民众意愿。这种部分搬迁安置的方法有利于城市活力的保持与快速再造，也是长远发展的保证。

古镇东南部的水系拓展，马家港延伸至石瀚经望津河、凤凰河入南市河，又南下竹园港、黄家荡、虹映港，接南浔南部主水脉笕五塘，修复与贯通了东南部水系，形成了平行于南市河的又一城市主纵线，嬗变为城市东部的结构水系，使水城有了更大的拓展维度（图17）。

作为历史上消失的水系，如后河、栲栳湾、洗粉兜等，基本是被填埋的道路，原则上以有利于城市复兴为前提，尽可能加以修复，通过多个支流河道的勾连使整个古镇水网更加发达，与城市居民生活更加密切，真正形成江南水乡的当代水城（图18）。

此外，南浔保存了众多古桥、古河埠，如通津桥、洪济桥，桥拱高达7米，南浔十景之"頔塘帆影"指的就是当年货船不下桅杆通过拱桥（图19）的

图16

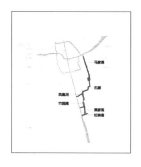  

图17（左），图18（中），图19（右）

情景，货船直出分水墩，销往上海，丝绸贸易盛况可见一斑。众多颇具特色的尽端是河道，被称为"兜"，也是曾经的私家码头，足见当年水运之盛。这些特色遗存都将进入水系修复的视野，共同推进水城再造。

七、水陆交通一体化构想

　　当南浔1.9平方千米的水系修复形成"申字加一纵"的一个基本的网状结构，构成古镇的城市架构，水路通达各区，双向通行，水上交通的规划自然呈现。水上巴士，一种清洁能源的交通，足以增加旅客体验乐趣，也为居民出行提供了更多选择。

　　历史保护区内计划修复的河道全长约8千米，假设电动船速8千米/小时、全线设12个码头，要求满足每班次相距5分钟（包括停泊），则单向正好投入船12艘，双向则需24艘。若留有余地，在高峰时段实行班次时间缩短一半，即平均2.5分钟一趟，48艘客船即可满足假设的需求.若每船平均20人，则每小时运力可达960人次，按每天7小时计，运力近七千人次，超过小镇居民量的1/3，理论上满足了水上客运的需求。而实际运力与船只数量，则完全可以根据届时的情况调整，并可用大小船平衡时段运量。

　　水运的实现将大大增加体验的乐趣，但要实现水运作为居民主要出行方式，仍需最大限度提供全方位的便利。如在古镇重要节点增加公交换乘，并贯达全区，再辅助以微公交、公交自行车系统；同时，规划好院落式微公交停车库（图20），充电、停车、租赁乃至换乘一体化，逐步形成居民与游客水上出行常态化；更进一步，古镇外围水上快运线，还将古镇与周边乡村连结成一个整体，将古镇与未来高铁站形成直达快线。这样，水上清洁能源巴士、换乘公交、租赁微公交、自行车，形成一卡通古镇公交系统。这将有效控制客车数量，减少道路规模，形成外环主路与渗透支路两级道路系统（图21），尽量保持水系的独立完整。无论对可持续发展还是新水乡体验都大有裨益。期望在这个依赖车行的快速时代，找到一种慢生活选择的可能，找回水生活的悠然乐趣。

八、水城再造之其他策略

　　如果说水系修复确定了水乡古镇的城市空间骨架，推演出以"水街—街巷"线型空间为组织的城市肌理与空间特征，那么，水陆交通一体化将是疏通城市气脉、有机运行的重要举措，为江南六镇最具特色的当代水城营造奠定了基础。

　　接着是历史遗产活化的"都市水乡"策略，首先是基础设施建设，以及合理配置社会公共资源。水系

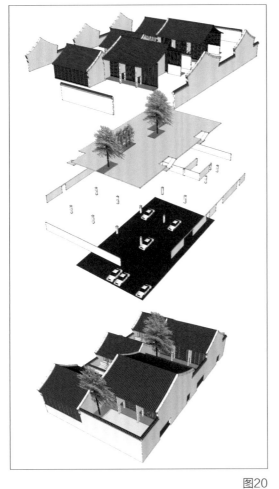

图20

图21

修复在确定城市主次街巷系统后,也为项目的具体实施指定了路线与走向。接下来是古镇保护利用的"文脉延续"营造策略,既要防止新特色对文脉基因的覆盖,更要避免过于表面化复制成为肤浅的传统样式"赝品"。当代性需要建立一种有"痕迹"的历史延续,并要求呈现城市活力,一种传统脉理与形式的新诠释。因此,具体落实将在"院落社区"与"新浔式建筑"两个层面给予回应。

南浔作为儒商精神集中体现的文化重镇,转型为真正意义上的文博小镇,在资源上是有绝对优势的,如以江南大宅门为背景,演绎高端文化产品的展示体验重地,推出系列文博活动,形成规模效应,或许称之为"文博大观"策略。这种文博资源的当代性转化途径也是一种产业振兴策略。而后,还有"乡贤共建"策略,鼓励南浔企业家与民间投资的投资,尝试一种乡村社会自治的痕迹,鼓励其以名仕乡贤的角色,共同参与魅力故乡的再造,培育浔腔、浔韵与水系生活美学,找回价值认同的终极幸福感。

从城市架构确立到古镇活力营造,从新兴文博产业振兴到魅力故乡再造,四种策略各有所指,与此伴随的是营造营运的品质化策略,它们共同作用,勾勒出南浔这座江南当代水城作为慢生活典范的魅力前景。

九、水城复兴活力再现

水系的修复，形式上延续了传统水网结构及庭院街巷的城市肌理，实质上，指明了基础设施的路径与方向，是至关重要的开端。以此为基础的都市生活配套与公共资源完善，将吸引新、老南浔人的居住与从业，尤其需要得到年轻人的场所与价值认同，形成新南浔人的回归，保持宜居宜业的活力，形成可持续发展的生活小镇，这是南浔城市复兴的基本目标。

当独一无二的当代江南水城体验叠合到厚厚的历史文脉中去，紧随而来也将是旅游的别开生面，一种居游两宜的"悠然庭院生活"自然受到广大文创人的青睐，"生活美学"魅力的文博小镇也将实至名归。

"小市千家聚水滨，轻舟日日往来频。桥从栅口分南北，南尽南新北北新。"今天，这首诗描绘的似乎不止过往的南浔，更像在叙述一直更替着、生长着、生活着的未来水城——南浔。

参考文献

[1] 周庆云撰，《南浔志》，江苏古籍出版社、上海书店、巴蜀书社，1922年
[2] 南浔镇志编纂委员会，《南浔镇志》，上海科技文献出版社，1995年
[3] 陈晓燕、包伟民著，《江南市镇》，同济大学出版社，2003年
[4] 断进、季松、王海宁著，《城镇空间解析—太湖流域古镇空间结构与形态》，中国建筑工业出版社，2002年
[5] 童寯著，《江南园林志》，中国建筑工业出版社，1984年
[6] 朱均珍主编，《南浔近代园林》，中国建筑工业出版社，2012年
[7] 徐迟著，《江南小镇》（长篇回忆录），长江文艺出版社，1993年

插图来源

图2、3、12《中国大运河申报世界遗产文本》，国家文物局；图4,网络；图6、7、8，马俊摄影、陆士虎著文，《南浔》，浙江摄影出版社；图10，周庆云《南浔志》；图9、13、16、19，作者自摄；图1、5、11、14、15、17、19、20、21，作者自绘。

鲁道夫·吕克曼
Rudolf Lückmann

德国安尔哈特应用技术大学副校长
建筑遗产保护专业教授、主任
亚琛工业大学建筑遗产保护专业博士
科隆大教堂保护项目建筑师（联合国世界遗产项目）
吕克曼建筑师事务所创建人和负责人
2010年　创建建筑遗产保护硕士专业

专著
《当代建筑设计手册》第39版，科兴 2011
《建筑设备》，科兴 2010
《被动式节能建造大全》，科兴 2011
《建筑设计大全》，科兴 2013
《木建筑大全》，科兴 2014

Prof at Anhalt University, Dean, Vice-president foreign affairs
Architecture at RWTH Aachen
Phd in Monumantal heritage RWTH Aachen
Architect Cologne Cathedral (UNESCO world monument heritage)
Diocesan-master Württemberg/Stuttgart
Founder and leader Architecture office Lückmann
2010 founder of the master-course Monumental heritage, director

Works
Aktuelles Handbauch der Baukonstruktion, 39. Aufl., Kissing 2010.
Technische Gebäudeausrüstung, Kissing 2010.
Passivhausatlas, Kissing 2011.
Hochbauatlas, Kissing 2013.
Holzbauatlas, Kissing 2014.

德国的村镇更新

鲁道夫·吕克曼

一、术语

村镇更新是一个由德国政府资助的项目，其目的是改善及提升村镇的各个方面。早在16世纪，德国就出现了现代村镇改造的模式。尤其自1950年以来，德国开始施行了一种较新的村镇改造模式，并且这个模式得到了政府的推崇。

由于二战以后的德国正处在战后的动荡之中，国家被分裂，在1200万的难民之中有15%以上的难民会融入当地居民的生活。当东德着眼于发展城市的同时，实行资本主义的西德却尝试强化村镇并为它们赋予了新的及多样化的功能。因此，西德实行了村镇更新这一项目。

村镇更新项目的核心不仅需要显著的结构和设计，在更新的过程中也必须将生态、经济、社会及文化因素考虑进去。当然，当地的市民也必须参与到这项更新的设计环节中去。

"村镇更新"这一术语指的是根据村镇定居点的位置及环境来进行量身定制的设计及发展，为的是可以纠正缺陷及弥补不足之处，从而改善人民的生活条件及环境状况。

当然，城市的重建也会面临与村镇更新相类似的问题，所以政府部门也必须为做好这一项目而实行一系列相关的政策及措施。由于这些政策与法规太过宽泛，所以在这篇文章中大多不会被提及。但是，建筑规范法（Baugesetzbuch BauGB），其作为德国建筑行业的基本法，其中所提到的关于城市发展及重组的内容将会做具体阐述。

二、历史背景

村镇更新虽然在当今社会受到广泛关注，但它却是基于一个非常古老的想法。它的出现是围绕着怎样能使村镇的发展发挥更大的作用。这也关系到19世纪时德国人对农业有了新的认识。这个过程即是农场合并及进行集约化发展管理。

农场合并及实行集约化的管理模式作为早期进行村镇更新的前提之一，在德国被称为"Kemptener Vereinödungen"。这两项举措完成于16至19世纪。在那段时期，许多农场被重新定位为生产专门的农业品种。这项举措不仅提升了当地农场的生产结构，还解放了一部分生产力。这个过程在当时为了应对人们接连出现的健康问题来说是必要的，尤其是在受到伤寒与霍乱的威胁下，农业还能获得丰收。

在19世纪初期又出现了不同的画面。我们处在浪漫主义的时代。一方面，城市化现象开始慢慢显现，但在另一方面又出现了人类回归自然的趋势。这也使得国家美化运动应运而生，并提出了更为全面的观点：在"更新"与美化周围环境的同时也要对农业（合并，后来被称为土地兼并）、园艺（后来的园林建筑）及建筑（后来的城市与村镇更新）进行整改。

这项运动在当时有许多非常有名的追随者，例如利奥波德三世王子。Friedrich Franz of Anhalt-Dessau (1740—1817), Prince Hermann von Pückler-Muskau (1785—1871)，而提到有影响力的精神代表，不是别人，正是Johann Wolfgang von Goethe (1749—1832)。这些可以从他的作品"亲和力"看出——是选择做实用的理想主义者还是做更好的理想从业者，例如巴伐利亚的建筑从业人员Gustav A. Vorherr (1778—1847) 在其俱乐部的杂志上写的《关于在巴伐利亚土地美化的建筑月刊》。在文章中，他提出了具体改善村镇的想法并更有效利用了过道。

但是，这项运动并未能阻止农业工业化的进程，人们更关注于时间与机器。由于人们对这些开放领域的编辑与设计的热衷，在接下来的很长一段时间村镇更新设计与改造被推入了幕后。尽管仍然有一些非常大胆的试验，特别是普鲁士地区的土地整治管理部门，这些部门帮助该地区调整村镇结构并将土地所有制发挥得更加有效。

在20世纪50年代的后半叶，村镇更新又重新回到公众的视线。农业专家提出实现走廊地区的现代化是远远不够的——在农业结构转型的时期，村镇也必须进行结构调整和实现现代化。

全德国上下都响应了关于进行村镇更新的呼声：在1950年后期，农业部作为主要负责部门批准进行村镇更新的项目。而进行村镇更新项目的主要发起部门则是农业协会。其中之一便是农业社团（ASG）。Tassilo Tröscher (1902—2003)，时任黑森州的农林部部长（1967—1970），他牵头当地财团，合力改善并提升了黑森地区的农业结构。这些社会团体便是在20世纪五六十年代力推农业改革及村镇更新的发起组织之一。

农业改革及村镇更新的根本目的是实现村镇的现代化。特别是，当时的人们尝试淘汰农业工艺的剩余物。除此之外，人民并不能有较高的生活质量，虽然为了能让年轻人留在村镇而提供了住房，但是却没有良好的卫生条件及社会生活（关键字：乡村社区）。在许多农户看来，村镇更新必须依赖于农业经济的发展及农业补贴。

但是很快人们就意识到村镇更新仅仅依靠单一的、随机性的资金是不够的，它所需要的是一个全面的资金体系来支持整个项目的完成。因此，自1965年开始，所有的村镇更新项目都要求有完整的资金体系，并且村镇发展规划筹集资金的一项重要条件。

为了能更好地进行城市与村镇的更新，联邦政府还进行了深入研究并启动了试点项目，还在1971年出台了城市开发法规。村镇发展规划受到了民众的欢迎并深得人心。因此，村镇发展规划的示范项目成为了当时的一个重要新发现。

土地兼并管理于1970年加入到了村镇更新的项目中。政府颁布了相关法令要求在每一次的土地整治中，都要提出一个相关的村镇更新规划。更好的是，这项举措也成为了城市发展法规的基础，并且正式地运用到城市及村镇发展中去。

但是，村镇改造与村镇更新这两个项目在当时分属于一个部委下的两个行政部门所管辖，所以导致了在处理某些事情上存在一定的分歧。在1973年，当完善了农村的基础建设和调整了土地结构后，这场对立也随之结束了，两个行政部门也被合并。

除此之外，政府也贯彻实行了一些相应的措施。但是，自1971年城市发展基金实行以来，将近全部的资金被用到城市中，这也使乡村代表感到失望——因为一些一流的村镇失去了乡村原有的风味。

三、村镇更新中的土地兼并

1976年，德国联邦议院同土地整治管理的相关部门决定对土地兼并法案做出修正。在这项特殊的法案中，"村镇更新"这项条目被首次提出，并且它们被嵌入到了"推动农村发展"这个领域的任务中。

村镇更新这个想法也因修正法案而重新展现在大众面前，尤其是国家美化运动的复兴及德国对村镇更新将采取实际行动——这些都归土地整治管理的相关部门进行管辖。而村镇更新项目也会依据土地兼并法案来实施。

当然，有些保守主义者及建筑师、园艺设计师都担心土地整治管理的相关部门不会扩展及深入村镇更新任务，并认为这些部门只会对一些不敏感的村镇地区进行更新改造。因此，那些由Dieter Wieland (born

1937)设计的村镇改造案例常被人们讽刺为非自然"漂亮"的村庄。

为此，他还使用了"我们美丽的村镇"这个比赛中的范例。但是Wieland也在他的"我们丑陋的村镇——1975欧洲建筑遗产年的一大贡献"的展览中展出了一些应该被制止的村镇更新的案例。这种整修方案自1960年以来在Baden-Württemberg地区却被当成了村镇改造及更新的范本。这篇报告中提到了这种村镇更新的方法则是将老的建筑结构进行全部拆除，而这种行为被视作是非常不好的。

在村镇更新中涉及土地兼并后，一些建筑师与当地保护主义者在村镇更新规划上从冲突中寻求信任合作。有名望的建筑师对于管理这项工作可以起到帮助。他可以运用自己专业的经验来调解和平衡各方的意见，而这项过程在村镇更新中是必要的。

然而，由于村民及规划专家因为思想与兴趣的不同而常常产生对立的观点。例如处在巴伐利亚林区的Hufendorf Kreuzberg，由于这个地区的村子是以放射状分布，最终建成的"循环路"起初被搁置多年，对整个村镇进行更新的设想也受到抵制，这个提案更遭到了保护主义者的强烈反对，但是毫无效果。

相反的，也有一些比较好的项目涌现。例如，the gentle village renewals这项作品获得了Klosterdorf Niederaltaich奖项（distr. Deggendorf），而in the secularized monastery courtyard of Polling (distr. Weilheim-Schongau)这项作品获得了专业领域的认同。

总体来说，最初村镇更新改造的要求是强调地区之间的差异。由于存在不同种类的农场，发展条件也不均衡，在海峡地区存在着大量的军事设施残垣，许多以小农结构为主的村镇是村镇更新的主体，所以在实行建筑和土地改造的过程中会产生强烈的反应。

大部分地区都进行了大规模的农业结构调整。当村镇更新及改造步入结构调整及大力发展基础建设的阶段时，那些当初没有进行农业结构调整的地区也随之跟上了步伐。这就到了村镇更新的后期阶段。

早期村镇更新改造项目能够迅速且广泛被大众接受是因为在1978年—1981年实行的未来投资项目（ZIP），在投资项目中所筹集到的资金被投入到了村镇改造中去。

地区变革及规划，完善法律法规，土地整合，大力建设基础设施及实现住宅及商用建筑的现代化是为了提高农业效益和提升整体经济。若没有财政的支持，村镇更新改造将寸步难行。

四、村镇更新的新历史

1981年在ZIP期满之后，因为村镇更新基金被重新规划，所以联邦政府不得不拒绝各州想进行村镇改造的要求。高额的负债迫使总理Helmut Kohl进行财政储蓄，但这些资金却常常被使用在了错误的地方。

3年之后——同样处于Kohl执政期间，联邦政府农业部长Ignaz Kiechle(CSU 1930—2003,自1983年至1993年任联邦食品、农林业部部长)纠正了这一决定，将改善农业结构及沿海保护这两个项目提升为联邦任务。

"资金使用指导规则"及村镇更新必须满足当地的需求并要与经济实力达到平衡，这有助于那些经济实力不足的地区。这些地区对这项计划表示非常欢迎，因为实际上他们为能进行村镇更新的请求已等了三年之久。

但是，这项指导规则更侧重于农业领域，这也是为什么一些混合生产结构的地区并不推崇这项规则的原因。在各联邦州进行的这项过程不尽相同，在Bayovaria, Baden-Württemberg及Hessen地区，当地的行政部门可以更进一步，因为他们可以借鉴当地历史中早期村镇更新的案例与经验。

巴伐利亚自由州并不喜欢被忽视。为回应联邦政府的拒绝，巴伐利亚议会在1981年的5月19日通过了关于自主进行巴伐利亚州村镇更新项目的决定。在1982年，这个提案被呈交到了当地的议会。随后，巴伐利亚州政府又对这个提案扩充了内容并将它扩展为一个整体的社区开发规划。

这项尝试为正在进行的农业结构化调整提供了答案。一方面，有越来越多的农民不得不放弃他们赖以生存的生产模式，大量的生产场地遭到废弃。而另一方面，也有许多农民需要更大的农庄，更大的厂房，更宽的道路去解决剩余劳动力的问题。

此外，商业慢慢成为了发展的核心，一些手工业退出了当地的供应市场，不再有人选择村镇作为居住点。因此，这也相应地对人口结构，经济发展和村庄及自然景观产生了一定的影响。

因此，巴伐利亚州的村镇更新项目又往前走了一步。而其他地区因为要在联邦政府的指导下才能获得改造资金，所以他们的村镇更新项目便落后了。从巴伐利亚项目我们可以得出的结论包括有形及无形的方面，及规章和行动标准。

因此，一些古迹、地理环境和村镇历史，民俗，生态及社会文化等因素都应该在村镇更新规划中被考虑到。因为早期就适用了参与规划和决策的制度使得巴伐利亚州直到今天在进行村镇更新规划时都会考虑地区政治环境、人口结构等因素。

有趣的是，有证据显示对村镇更新投资可以在政府提高税收时带来财政回报。所以，村镇发展是一项很好的投资项目。在政府提供的资金中，包括私人资金给与，据在慕尼黑的ifo研究所多次引述并反复确认其研究成果显示，村镇更新能带来财政回报。对村镇更新的投资引发的乘数效应使得资金提高了七倍多。这指的是对私人，自治及其他公共区域每投资一欧元，就可以引起其他6欧元的效应。

在人口的影响下，今天德国的村镇更新开始朝着能源化的方向过渡。德国将进入老龄化的社会。由于持续的移民潮以及低生育率造成了年轻人的减少，这些现实将会使周围环境产生巨大的问题。因此，保持年轻家庭的增长是很重要的。今日我们面临的新任务是要以年龄作为设计导向有针对性地进行村镇规划。

在另一方面，村镇也应做好应对大量人口涌入的准备。各个村镇社区必须欢迎新市民的到来。在村镇更新的过程中不仅要做到仔细规划，而且也要大力建设及完善家庭式的基础设施。

五、村镇更新——展望未来

村镇更新让市民对未来的生活产生信心。因为他们全程参与了更新改造的过程，如参与规划，达成共识并帮助改造。这也证实了德国各州关于村镇更新进行的所有社会科学的研究。

一些积极的市民比一般市民更能在未知的未来及不确定中意识到及肯定事物的轮廓。这个问题将是怎样去挖掘我们身上"沉睡"的潜能，我们是否能够促使这些潜能去承担责任——而村镇更新给予了我们这个机会。

但这个机会也不是自行产生的。一方面，国家必须出台一系列的措施来支持并帮助国民。为了能更好地进行农村改革发展，除去资金分配，行政部门还应该提供强有力的专业支持。政府人员及一些专业人士会一起准备并介绍各方最新的科学发现，并且关于村镇更新的研究结果将会付诸于实践。

另一方面，自治区、村镇及农户可以从一系列的服务项目及机会中获得帮助，这些辅助措施——通常被称为合作及讨论。一些社会团体，政党及社区的知名人士有责任去推动村镇更新及改造。

Village Renewal in Germany

Rudolf Lückmann

1. Term

Village renewal is a government-sponsored program to improve various aspects of village community. First forms of modern village renewal can be detected in Germany already in the 16th century. The village renewal newer type is carried out in particular since the 1950s and promoted by the state government.

The causes lay in the lost World War with the upheavals that resulted from it. By 12 million refugees around 15% more people had to be integrated into the community on the remaining territory. While the GDR put their eye on cities, attempts were made in the capitalist West Germany to strengthen the villages and give them new, viable functions. Here programs of village renewal were launched.

Core of the village renewal program are notably structural and design, ecological and economic and social and cultural support measures for the participating villages and communities. Governments in Germany had ot learn, that this kind of processes must be designed with the citizens together.

Under the technical term "village renewal" is meant measures to order, design and development of village settlements, which aim "to remedy existing deficiencies and irregularities, thereby improving the living conditions of people and the environmental situation".

But of course, there were similar problems in the cities there the authorities had to act a professionally related urban action for urban regeneration also.. This sometimes even more extensive items go unnoticed in this article. Basis of all interventions is in Germany, the Building Code Law (Baugesetzbuch BauGB), where the urban development restructuring is specifically addressed.

2. History

In this village renewal is not only a concern of the present; rather it is based on a very old idea. It revolved around to make the village more effective. This related to the 19th century in particular to the new understanding of agriculture in Germany. The processes were concentrations and mergers.

As one of the earliest precursors of the village renewal were the so-called. "Kemptener Vereinödungen" in Germany. They were settled in the 16 to 19 century. During these centuries many farms were resettled in the field layers. This should improve the local structures and make them more loosen. This procedure was necessary because occurred through the increase of population health problems were parallel growing. In particular, the typhoid and cholera had rich harvest at the too narrow structures.

In the early 19th century a different picture emerged. We are in the age of Romanticism. On one hand the urbanization began slowly, there waso on the other hand a trend that could be called back to nature.

It followed to country's beautification movements with a holistic view: to "renewal" or beautification of the surrounding being with actions of agriculture (merger, later land consolidation), the horticulture (later landscape architecture) and the architecture (later Urban and Village Renewal).

The movement had such illustrious followers as Prince Leopold III. Friedrich Franz of Anhalt-Dessau (1740-1817), Prince Hermann von Pückler-Muskau (1785-1871) or influential spiritual representatives as none other than Johann Wolfgang von Goethe (1749-1832). It can be seen in his education lyric his "Elective Affinities"- Practical idealists or better idealistic practitioners such. as the Bavarian construction officer Gustav A. Vorherr (1778-1847) wrote in the own club magazine "monthly Journal of Construction about the land beautification in Bavaria". He designed very concrete ideas for improving the village shape and more efficient use the hallways.

However, this movement could not prevail against the industrialization of the agriculture itself; concern about the time and machine-oriented. The editing and designing of open fields pushed the design and renovation of the villages for a long time in the background Although there are always some courageous experiments especially of the Prussian land consolidation authorities, to help shape the village and make at least to land tenure more effectiv.

Upturn widely learned the village renewal again in the second half of the 1950s. Agricultural experts noted that the modernization of the corridors was not enough - the villages had to also be adapted and modernized in the agricultural structural change.

Everywhere in Germany found this requirement echoes: From the end of the fifties the go-ahead to the so-called village improvements in responsibility or of the agricultural administration or ministries. Main initiators were nationally active agricultural associations.

One was the Agricultural Social Society (ASG)- The Hessian Minister Tassilo Tröscher (1902-2003), Minister of Agriculture and Forestry in Hessen 1967-1970) led consortium to improve the agricultural structures in Hessen (AVA). These organizations are among the initiators in the 1950s and 1960s.

The aim was to make the village modern. In particular, an attempt was made to eliminate agro-technological residues. In addition, shortcomings in the areas of quality of life and to remain willingness of youth formative should housing, hygiene and social life (keyword: Village community house) be reduced. According to this agrarian and geared to farmers and farming families view the village renovation was heavily geared to agrarian-economic and agro-social subsidies.

Very soon realized it that a more comprehensive and especially methodologic approach instead of a pure, sometimes quite randomly appearing individual funding was necessary was therefore required from 1965 a so-called. Village development plan as a condition of funding.

The federal government had launched studies and pilot projects for urban and village renewal. She wanted to experience the outstanding urban development law of 1971. The village development plan was well received and was suddenly in everyone's mind. As a result of the demonstration projects of the village development plan was a key new finding.

The land consolidation management joined 1970 this development. They decreed that in every land consolidation process, a village renewal plan should be created. At best, this should be the basis for the expected of her windfall of urban development law, which officially should apply to cities and villages.

Village renovation and village renewal, however, were in two administrations within one Ministry. This led to everything else than counter together. These controversies ended in 1973 when the agricultural village improvements with the strong infrastructure and land readjustment were established. Aligned village renewal planning of land consolidation management was merged.

In addition, there have been some means of implementation. Of the urban development funds available since 1971, however, nearly all feed millions in the cities and went to the disappointment of rural representatives - by in some flagship villages fell for rural almost nothing from.

3. Land consolidation through village renewal

1976 the German Bundestag decided with considerable contribution of the German land consolidation authorities an amendment to the Land Consolidation Act, in a specialized law were the word "village renewal" was called first time. They were embedded in the task of the "promoting rural development" of areas.

This was the rebirth of the former village renewal ideas in particular the country's beautification movement and the actual start of the renovation of villages in Germany - under the responsibility of the land consolidation authorities. The village renewal was usually embedded in proceedings under the Land Consolidation Act.

Of course there were first at home- and conservationists as well as architects and landscape designers who fears that the land consolidation authorities had not grown to this task. They assumed that would handle insensitive to the villages. The doomsday scenario were installed by Dieter Wieland (born 1937) - often caricatured with unnaturally "pretty" villages.

To this end, he used examples from the competition "Our village is beautiful". But Wieland showed deterrent examples in his show "Our village is ugly - A contribution to the European Architectural Heritage Year 1975" The area renovations appeared in some areas of Baden-Württemberg were model village renewal from the 1960s. were done with evil actions. The report shows village which renewal was almost the total demolition of the old building structure as a result.

But after the land consolidation was involved from the beginning, the architects and preservationists in the village renewal plans found to a trusting cooperation from an impending conflict. That required a management in which often distinguished architects could help. Her professional experience in mediating and balancing different intentions, what is needed in such a process.

Nevertheless, due to the often contradictory ways of thinking and interests of villagers and experts was not always popular and with general pleasure. An ultmately built "circle road" in the radial village of Hufendorf Kreuzberg in the Bavarian Forest burdened for years the image of the village renewal down. The proposal was opposed by preservationists vehemently to no avail.

Conversely, there are also positive projects. For example, the gentle village renewals winning Klosterdorf Niederaltaich (distr. Deggendorf) or in the secularized monastery courtyard of Polling (distr. Weilheim-Schongau) lifted the acceptance of village renewal decisively in professional circles.

Overall, there was initially a strong regional disparity in the village renewal demand: Due to the difficult kind of farmers yards and development conditions and the large infrastructural residue in the narrow, many of smallholder structure embossed pile villages was the village renewal rather dominant. It was very strong as a building and land readjustment program applied.

The parts of the country with large agricultural structures were talking initially sharply - apart from a few exceptions. They followed only later, when the village renewal stepped on the structural and infrastructural phase. That was one of the later stages of village renewal.

The rapid and wide acceptance of village renewal in the decisive breakthrough for the early years was probably ultimately possible only because in the course of the Future Investment Program (ZIP) 1978-1981 a warm windfall has also been provided for the renovation of villages. A simultaneous, settlement and landscape enclosing village renewal and land consolidation allowed expedient

Country-changing or new-designing, legal regulations, consolidations, infrastructures and modernization of residential and commercial buildings are often done for the benefit of agriculture and the general economy. Without financial support to programs of village renewal, this cannot be implemented.

4. A new village renewal history

After the expiry ZIP 1981, the federal government was closed to the demands of countries according sequel or remake of village renewal funding. The high indebtedness forced the government of chancellor Helmut Kohl to do savings, which were often carried out at the wrong place.

Three years later - also during the times of the Kohl government the Federal Agriculture Minister Ignaz Kiechle (CSU 1930-2003, Federal Minister of Food, Agriculture and Forestry from 1983 to 1993, corrected this decision. It resulted in the joint task of improving agricultural structure and coastal protection (GAK) as a federal level.

The "funding guideline" and village renewal measured the subject in the balance of needs and economic strength. . This helped especially the financially weaker countries. They already welcomed the program because they actually had to wait for the resumption of strong demand village renewal in their own country for three years.

However, the guidelines were very closely held and emphasized agrarian areas why they were little appreciated by many regions with mixed structures. To commence the process in the various federal states differed widely under way. In Bayovaria, Baden-Württemberg and Hessen the administrations were further, because they could look on early village renewal programs in their history.

This neglect the Free State of Bavaria could not like. In response to the refusal of the Federal Government in Bonn it was decided by the Bavarian Parliament of 19 May 1981 towards the establishment of an autonomous Bavarian village renewal program. 1982 it was presented to their parliament. Subsequently, the authorities' worked it out to content and expanded it into a holistic community development program.

This attempted answer had to be given to the ongoing structural changes in the agriculture. There was one hand with increasingly farmers, which gave up their business, with disused site picture formative buildings. On the other hand there were remaining strong farmers with need for larger farmsteads, larger buildings, partial or full relocation, wider roads etc.

Also, the commercial and the commercial structure around had to take care of the development. The craft withdrew from the local supply. No one could live from his village anymore. So this was corresponding with an impact on population composition, economic and villages and landscape.

Thus the Bavarian village renewal program was taken further content. Financed under federal guidelines the village renewal programs of other countries lagged behind. The Bavarian program concluded and includes both tangible and intangible aspects, disciplines and actions.

Thus were monuments and home care movements, settlement and village history, folklore, village ecological and socio-cultural aspects as an integral part of the village renewal plan. By early application of participatory planning and decision methods won the Bavarian village renewal until today a continuing high level of awareness in state and local politics, and in the population.

Interesting is the evidence that the investment in villages renew has a financial return for the governments in the form of higher tax revenues. Village development is a good investment. The funding the government provides, including private funding bring, according to a much quoted and repeatedly confirmed research results of the ifo Institute in Munich the money back several times. The investive multiplier effect in the village renewal stands in relation to the funds up to seven times. It means that each Euro inserted will trigger another six Euros from those investments in the private, municipal and other public areas.

Today the village renewal in Germany is dedicated to the energy transition, but especially the demographic impact. German will be overaged. The ongoing emigration or falling birth rate and the resulting lack of young generation faces in structurally weak peripheral region serious difficulties. It is important to keep young families. New tasks of internal development and a targeted age-appropriate village design must be touched.

On the other hand villages have to struggle for social integration in excellent of population influx. They have to welcome the new citizens. Careful planning developments or family-style infrastructure should accompany the process.

5. Village Renewal - opportunity for the future

Village renewal provides for the citizens confidence in the future. You want to have a say responsible, involved in planning, co-decide and help shape, This confirmed all relevant social science studies for village renewal in the German states.

Active citizens are therefore more able to tolerate the unknown future and uncertainties to sense yes and shape. The questions will be how we bundle the "sleeping" potential of our people. Will we be able to move them from enduring to bear responsibility? Village renewal offers the chance.

This does not happen by itself: On the one hand, the state must support its population to help themselves. To this end, in addition to the allocation of funds includes the strong professional help of an administration for the rural development. It's partners in the government and free profession belong together preparing and introducing each of the latest scientific findings and research results about village to help them into practice.

On the other hand municipalities, villages and villagers have to use the services and the chance to help themselves subsidiary - usually claiming the team and discussion culture. Societal organizations, parties and dignitaries in the communities have a high responsibility to encourage this process.

沈杰
Shen Jie

浙江大学建筑学系教授、博士研究生导师
俄罗斯圣彼得堡国立建筑大学建筑学博士
美国麻省理工学院建筑系高级访问学者
中国建筑学会建筑师分会建筑技术专业委员会副主任委员
乌克兰国家建筑科学院外籍院士
俄罗斯国际生态、人类与自然保护科学院通讯院士

研究方向
可持续建筑设计与理论
生态城市与人居环境研究
俄罗斯文化艺术研究

在各种中外学术期刊和国际会议上发表了六十余篇学术论文,先后承担十余项省部级纵向课题和横向科研项目。

Professor, academic supervisor of doctoral candidates, Architecture Department, College of Civil Engineering and Architecture Zhejiang University
Doctor of Architecture, Saint-Petersburg State University of Architecture and Civil Engineering (SPSUACE)
Senior visiting scholar, Department of Architecture, Massachusetts Institute of Technology
Vice chairman, Architectural Technical Committee at Architects branch of Architectural Society of China
Foreign academician, Ukrainian Academy of Architecture
Correspondence academician, Russia International Academy of Ecology, Man and Natural Protection Sciences
Research interests: Sustainable architectural design and theory, ecological city and human settlement environment, Russian culture and art
Shen Jie has publicated over 60 academic papers in various academic journals and international conferences at both home and abroad with over 10 provincial or ministerial research projects.

建筑遗产的"可利用"保护策略
——以广西鼓鸣寨滨水区保护与更新为例

沈杰　张蕾

【摘要】建筑是人类文明的载体，建筑的发展标志着人类文明的进程。从古至今，在世界各地，建筑无不被视为代表人类文明的标志性成就之一，因此，建筑遗产的有效保护和适宜利用具有深远的意义。文章将分析建筑遗产保护的意义作为切入点，阐述了采取有效保护和适宜利用，即"可利用"的保护方法的必要性，并进一步提出"可利用"保护的原则为"根植于过去，立足于当代，放眼于未来"，提出当代的利用与保护是为了在过去与未来之间搭建一座沟通的桥梁，实现文化的历史传承与文明的多样发展。最后，结合广西鼓鸣寨的具体案例，对以上原则和方法进行剖析，通过鼓鸣寨的"可利用"保护设计，我们努力尝试寻求一条肩负着衔接过去文明与当代文明的、通向可持续未来的建筑遗产保护发展道路。

[Abstract] Architectures are the carriers of human civilization, and the development of architectures symbolizes the progress of human civilization. In all ages, architectures have been regarded invariably as one of symbolic achievements of human civilization all over the world. Therefore, the effective protection and proper utilization of the architectural heritage is of far-reaching significance. We take the analysis on significance of architectural heritage protection as an entry point to expound the necessity with effective protection and proper utilization, namely "available" protection methods in this article. Moreover, we put forward the principle of "available" protection for the past foundation, current foothold and future expectation, and raise that the purpose of cotemporary utilization and protection is to build a communicative bridge between the past and the future in order to realize the historical inheritance of culture and diversified development of civilization. Finally, we analyze the above principles and methods combined with the concrete cases in Guangxi Guming Stockade and we try to find a development way of architectural heritage protection to link up with past civilization and contemporary civilization and lead to a sustainable future.

一、建筑遗产保护与利用的意义

建筑作为"历史的史书"承载着人类文明的脉络和人们生活的空间，其中留存下来的建筑遗产更是蕴含了丰富的文化内涵和历史遗存，因此建筑遗产的有效保护和适宜利用具有深远的意义。

1. 作为文化载体，建筑的发展标志着人类文明的进程

从古至今，在世界各地，建筑无不视为代表人类文明的里程碑。法国文豪维克多·雨果认为："人类没有任何一种思想不被建筑艺术写在石头上……人类的全部思想，在这本书和它的纪念碑上都有其光辉的一面。"[1]；中国建筑历史学家和建筑教育家梁思成先生指出，"建筑是一本石头的史书，它忠实地反映着一定社会的政治、经济、思想、文化"[2]；美国著名艺术史家H.W.詹森在其著作《詹森艺术史》中写道："当我们想起任何一种重要的文明的时候，我们有一种习惯，就是用伟大的建筑来代表它"[3]。

因此，建筑遗产的有效保护是人类文明历程保护的重要一环。

2. 作为研究载体，建筑遗产是历史研究和学术研究的直观范本

建筑遗产反映了过去的自然条件和气候条件，反映了当时的宗教、文化和意识形态，反映了历史的经济和技术发展。

在建筑历史研究中，最主要的两种资料为实物资料与文献资料。文献资料是了解建筑历史的重要信息来

源，而作为实物资料的建筑遗产因为更加直观与客观，因此其资料的可信度要更胜一筹，可以用来验证文献资料记载的真实性、纠正文献记载的错误，甚至可以填补文献资料的空白。

建筑遗产的有效保护可以为专业学者们的历史与专业研究提供数量更多、质量更佳的第一手直观资料。

3. 作为功能载体，建筑遗产的适宜利用可以沟通过去与未来

建筑遗产的特点之一，是其内部的建筑空间仍然具备使用的功能价值，因此在保护的基础上适宜地对其利用，不仅是资源的循环利用，同时也使建筑遗产在承载着历史的厚重文明的同时，融入到当今社会之中，并在未来继续提供使用价值。

对建筑遗产的适宜利用可以为过去和未来搭建一座沟通的桥梁。

4. 作为实物载体，建筑遗产的适宜利用可以促进可持续发展

建筑遗产不仅呈现出地域分布的差异性和国家民族的多样性，同时也展现了历史的延续性、传统的继承性和时代的特征性。

《21世纪议程》[4]指出，可持续发展的一个重要方面是保护历史文化遗产。在保护的基础上对建筑遗产的适宜利用是文明延续的重要载体，一方面避免文明成为无本之木，无源之水；另一方面使得特定的建筑文明得以留存并延续，免于湮灭。延续建筑文明的多样性，这在全球化日益全面和深入的今天意义尤为重大。

综上，建筑遗产作为文明延续的载体，其价值需要通过对其有效的保护和在保护基础上的适宜利用得以发挥作用。有效保护与适宜利用，即是"可利用"的保护。

二、"可利用"的保护原则——过去—现在—未来

如何发挥建筑遗产的作用，有效地进行文明和历史的传承，这是建筑遗产保护的核心问题。我们的回答是"根植于过去，立足于当代，放眼于未来"。这是我们进行建筑遗产保护时的总原则。我们的目标是将有效的保护与适宜的利用有机结合，实现建筑遗产"可利用"的保护。在广西鼓鸣寨村落保护与更新的方案设计中，所有的工作都在上述总原则的指导下进行。

鼓鸣寨，位于中国广西，坐落高山盆地，前临平湖，依山逐级而建。全村民居大部分为清代及民国时期修建，均为夯土建筑，原生质朴，是一处环境优美、交通闭塞的独守了数百年的壮族村落建筑遗产[5]。（图1）

1. 根植于过去

鼓鸣寨至今保留下来的老房子，都是运用传统的夯筑工艺，使用原生的土、卵石、木头等建筑材料建构而成。其原有的村落空间和建筑空间表明，生活在这里的人过着日出而作、日落而息的农耕生活。由于经济的原因，并不能追求生活品质和室内外空间的物理环境（图2）因此，在鼓鸣寨滨水区的保护与更新设计中，重点之一就是如何实现夯土技艺的表达与传承及对滨水区原有空间脉络的梳理与再现。

2. 立足于当代

滨水区保护与更新之后，将作为当地一项重点的旅游开发项目面向现代社会的生活方式。因此在设计中，不能孤立地强调有效的保护，还要在保护的基础上做适宜的利用改造，改造后的建筑的内外部空间能够满足现代的生活需求。通过空间品质的提高来促发其间的物质生活与精神生活品质的提升，并为生产和生活方式的转变提供适宜的空间，最终使建筑遗产重新焕发生机，适应当代生活。

图1

图2

3.着眼于未来

经过"可利用"的保护改造，鼓鸣寨滨水区传统老旧的夯土建筑能够承载过去的厚重，承担当代的功能，"活"着走向未来。

三、"可利用"的保护目的 —— 价值挖掘

在进行"可利用"的保护设计时，通过三方面的价值挖掘体现上述原则与目标。

1.历史史料价值

建筑作为文明载体反映历史社会面貌及工程艺术成就，其本身的存在不仅是物化的历史体现，其中承载的生活更是动态的历史传承。在建筑遗产保护中，历史史料的有效传承是基础工作。

2.文化情感价值

建筑遗产所传达的地域性与民族性可以激发强烈的情感共鸣，引发感同身受的文化与情感认同。在建筑遗产保护中，以文化彰显为基础，显现出对特定历史文化传统的认同，使人产生强烈的文化情感的归属感。

3.物质功能价值

建筑遗产在成为"遗产"之前是"活"的建筑，有实际的使用功能。在成为遗产之后，反而有可能变成静态的单纯展示。从某种意义上说，这样的静态建筑遗产已经"死去"了。我们认为，维持建筑遗产的最好的方法是恰当地使用它们，而不是将其"隔离"起来，装在橱窗里，只可参观，不可参与。建筑遗产的特点决定其内部建筑空间的物质功能价值仍有很大的再利用可能，并可以通过这种再利用对建筑遗产重新赋予生机。

四、"可利用"的保护案例 —— 鼓鸣寨滨水区保护与更新设计

位于广西的鼓鸣寨，由于地处偏僻及传统的生产生活模式，民居逐渐凋敝，建筑日益破败，这处500年历史的村落正面临着与时代脱节的局面（图3）。为了避免建筑遗产的逐渐消失，我们对鼓鸣寨滨水区进行了研究与再设计。在设计中，秉持"根植于过去，立足于当代，放眼于未来"原则，发掘其历史史料、文化情感及物质功能三方面价值，在传承古村落的历史文化的同时，使鼓鸣寨建筑遗产重新焕发青春。

图3

1. 规划层面 —— 复合为面向多群体的综合功能景观带

 a.空间组织

 强调村落外部公共空间的优化，改变原有的混乱宅间空间与拥挤的交通组织。整合出以外部公共空间为核心的五个功能组团，既有静态展示，又有深度体验；既有工艺环节，又有生活环节；既有安逸的休闲区，又有动感的娱乐区。（图4）

 b.流线组织

 在原有道理基础上重新梳理交通，组织成七列纵向小路，紧密联系地块中原有的主要两条流线。（图5）

 c.室外空间梳理

 划分成五个功能组团，每个组团按照组团功能设置一个以上集中的室外空间，使得滨水带空间疏密有致，动静相间，室内空间与室外空间交替穿插在流线上。（图6）

2. 建筑单体层面 —— 最小化改动建筑风貌

图4（左上），图5（右上），图6（左下），图7（右下）

29

在最大化保留原有建筑立面和屋顶机理的前提下，对建筑的结构和设施进行改造，打造适应现代功能，舒适的内部空间环境（图7）。在建筑内部插入与原有夯土建筑脱开、完全独立的钢结构体系，并在夯土墙体易受破坏的应力集中的端面施加钢保护带（图8）；在原建筑山墙内侧用竹材设置管道夹层，建筑内所用的水管、电管等基础设施均在夹层内解决（图9）。

3. 材料层面——绿色环保

改建所用材料主要选用钢、玻璃和竹子，均为可循环利用绿色材料。

4. 工艺层面——动态传承

建筑遗产"可利用"的保护中，其重要的一环为传统建造工艺的传承。建筑是静态的，工艺却可以代代传承下去。因此不强调针对原有夯土建筑的"永固性"进行保护，而是由其自然更迭，这样每一代都会有需要重新夯筑的建筑，每一代都会有传承这门工艺的匠人师傅。通过这种方式，传统的夯筑技艺留存下来。

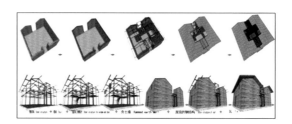

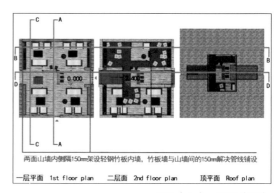

图8（上），图9（下）

五、小结

通过鼓鸣寨的"可利用"保护设计，我们努力尝试寻求一条肩负着衔接过去文明与当代文明的、通向可持续未来的建筑遗产保护发展道路。

注释

[1] [法] 维克多·雨果：《巴黎圣母院》。

[2] [美] 费慰梅著、成寒译：《林徽因与梁思成》，北京：法律出版社，2011年11月。

[3] [美] H.W.詹森著、艺术史组合翻译实验小组译：《詹森艺术史》，北京：世界图书出版公司北京分公司，2012年6月

[4] 《21世纪议程》是1992年6月3日至14日在巴西里约热内卢召开的联合国环境与发展大会通过的重要文件之一，是"世界范围内可持续发展行动计划"，它是前至21世纪在全球范围内各国政府、联合国组织、发展机构、非政府组织和独立团体在人类活动对环境产生影响的各个方面的综合的行动蓝图

[5] 文章中所用的鼓鸣寨照片均为"鼓鸣寨国际学生夯土建筑设计竞赛"官方提供。

亚历桑德鲁·梅里斯
Alessandro Melis

奥克兰大学高级讲师，教授建筑课程，包括可持续性设计，保护和建筑设计，同时也担任建筑技术系的主任，建筑学院博士后研究以及国家创新艺术研究院（奥克兰大学）的主管。专注于研究气候敏感设计，城区复兴，能量重新分配和保留。担任过维也纳应用技术大学、安特哈尔应用技术大学和萨萨里大学的客座教授和爱丁堡建筑学院的名誉研究员。应邀在纽约现代艺术博物馆和库伯联盟学院做过主题演讲。在罗马、佩鲁贾、惠灵顿(维多利亚大学)和克莱斯特（CPIT）和其他城市教授过课程。撰写了多部著作，并在欧洲和美国发表多篇论文。于1996年创办Heliopolis21，一家在意大利比萨，德国柏林和新西兰都有分支机构的国际公司。其项目在全球多地展出，并多次获奖，包括威尼斯双年展，并收录在国际期刊中，比如东京出版的《GA Documents》。2015年，梅里斯入选指导威尼斯双年展新西兰展馆。

As Senior Lecturer in Architecture at the University of Auckland, Dr Alessandro Melis teaches sustainable design, conservation and architecture design. He is also the head of the architecture technology department and director of the postgraduate studies (PhD) of the School of Architecture and Planning of the National Institute of Creative Arts (UoA). His specialist teaching, supervisory and research interests are in the fields of climate sensitive design, urban regeneration, energy retrofitting and conservation. He has been a guest professor in istitutions such as die Angewandte Vienna, the Anhalt University of Applied Science Dessau, and the University of Sassari. He has been honorary fellow at the Edinburgh School of Architecture. He has been invited as a key speker in institutions such as the Museum of Modern Art, New York, The Cooper Union, New York and lectured at the universities of Rome, Perugia, Wellington (Victoria) and Christchurch (CPIT) amongst other cities. He has written several books, journal articles and essays published in Europe and USA, also included in A rated journals. In 1996 he founded Heliopolis 21, an international firm based in Italy (Pisa), Germany (Berlin) and New Zealand (Auckland). H21 received a number of awards and its projects have been widely exhibited, including the Venice Biennale, and published in international magazines such as GA Documents (Tokyo). In 2015 Alessandro Melis was a finalist for the direction of the New Zealand

通过保护和旅游发展政策复兴托斯卡纳地区的古镇

亚历桑德鲁·梅里斯、奥鲁夫托·尔加土图依

介绍

目前对现存建筑的保护多是基于道德和环境原因,而不是出于保护历史遗迹和社会文化的需要。现在全球各国都应该减少过度开发包括土地在内的环境资源(Foley,2011)。在像托斯卡纳这样的高人口密度地区,人们应该将保护和"零"体积政策(Hermelink,2013)融入到一种系统的做法中去。该区域位于意大利的中北部,是全球UNESCO遗迹分布最密集的地区,这里是在都市和建筑级别实施先进的可持续保护措施的首选。人口向郊区的迁移以及边远村落的逐渐荒废对于历史建筑的保护非常重要,这促成了农村地区的旅游复兴。托斯卡纳地区已经成为了可持续性发展的典范,并同时成为了成功发展旅游的一个品牌,尤其是在盎格鲁-撒克逊国家(英国,美国,澳大利亚和新西兰)和欧洲中部以及北部地区。 基于发展旅游业设计的建筑保护方案已经成为当地政府和投资者就建设基础设施达成共识的一种途径,并已经促进了当地的发展。

背景

二氧化碳的排放主要(占全部的30%—40%)取决于建筑的设计、建造、维护和拆除(Ürge-Vorsatz, 2007; Mata, Kalagasidis, and Johnsson, 2013; Ibn-Mohammed, 2013; Ezema et al., 2015)。预计到2050年,全球人口会增长20%,而同时由于温室气体导致的全球变暖加剧,食品生产会缺少20%的土地。导致这一切的原因是海平面的上升,沙漠化和反馈现象。因此,作为可持续性领军组织的欧盟从20世纪80年代开始便推行零体积和零能量政策(Hermelink,2013)。 托斯卡纳这样的区域拥有重要的历史地位,境内遍布UNESCO遗址,且全球密度最高。在这里保护现存建筑是不二的选择,并能为经济发展提供帮助。

其模式非常简单,在托斯卡纳地区,荒废的历史村庄被改造为国际旅游目的地。其成功的原因主要是鲜明的当地文化特色(对于外国人来说很有吸引力),舒适的居住条件,以及整合了旅游和服务居民功能的新设施。 这确实是个"双赢"的策略,这些设施是当地政府和私人投资者谈判的工具。运动设施,会议室,图书馆和商店都为都市复兴和地区发展创造了条件,同时令人流连忘返。

个案研究

在Heliopolis21设计的诸多项目中,四个项目可以作为以上策略的典型案例:

Aulla项目是部分复兴策略实施的区域典范,其中包含用以替代独幢建筑的低密度建筑模式的有机复式都市发展模式,其将商业和运动设施整合到现有的都市中去,以供游客和当地居民使用。各具功能的古代村庄和新建村庄将生产和旅游整合到了一起(手工艺品,食品和酒,休闲以及健康等)。

地产发展商在Aulla资助下新建的体育中心是该项目的核心。它的主要作用是举行集会活动(游客,当地居民和国家运动队)。里面有一个活动大厅,一个水疗中心和各种活动操场,这里既有特定活动区域,又有适合开展室外活动的自然场所(散步和骑马等运动)。

高尔夫球场以及附属建筑(俱乐部,训练中心等)是Aulla和Castefalfi项目中的一部分。它的设计使用了最新的技术以最大程度减少用水。该设施促进了周边旅游业和社区群体活动的发展。Castelfalfi项目最初包括两个宾馆、一个会议中心、一个水上运动中心和诸多其他设施。古城堡周边村庄中的现存建筑将融合餐厅,商业和手工艺设施。

Nocchi项目的重点是并入Grazian古村落的18世纪建筑群的保护。 该项目旨在改变用于生产橄榄油的

老磨坊建筑。农庄和仓库将被用于存放农产品。人们共享公用室外空间以保持建筑物原本的完整统一性。建筑外部按照文献学保护研究结果（文献和档案研究）得以重新修饰,精心复原,修旧如旧。

  Libbiano的Nocchi项目内饰的灵感来自当代设计。 对其材料和"外壳"的原始结构的保护很清晰,现代材料和仪器意在满足当代使用者的需要。由于意大利地区的地质运动活跃,修缮工作的一个主要部分是专门整修建筑结构。地基的边缘和马路边缘都使用了钢筋混凝土加固,外墙周边也用混凝土进行了加固。地板和屋顶都用钢筋结构进行了加强,石质墙板中都嵌入了垂直的混凝土柱子。这些措施都增加了房屋的抗震性能。钢筋混凝土门框也加固了建筑上新开的出入口。 同时,由于墙体中加入了新的隔热层,减少了热传导,建筑的隔热和隔音效果也得以改善。

古代村庄的复兴

  复兴城市的重要性可以追溯到二战后欧洲工业的衰落时期(McDonald, Malys, and Maliene, 2009).。因为经济和其他原因,较发达区域和都市中心吸引着众多乡村居民前往。这就导致了边远的村庄和周边地区的荒废和荒漠化（Dinis, 2004）。由于当时关注的重点是都市发展所面临的挑战,古代村落的复兴并没有得到学者们的重视。 Galland等人在UNESCO最近发表的文章中指出,尽管全球各国以及大众媒体和投资者都在关注全球各处的遗产提名,但几乎没有人关心现存世界遗产的保护和管理。有文献指出乡村才能使得城市变得完整,因为城市本来就是乡村的分支。因此,当下有些古代村庄受到城市化的影响已经变成了城市郊区,复兴这些古代村庄就变得非常重要。Li等人 (2014)认为尽管四个阶段（20世纪50年代到80年代）之间没有明显的边界区分,它们已经可以解释都市复兴这个概念的来源。这些概念是都市重建(20世纪50年代)、都市再生(20世纪60年代)、 都市更新(20世纪70年代)、和都市再发展 (20世纪80年代)。这些概念都指向了复兴这个目的,它们在托斯卡纳的村庄中得到了阐释。

区域营销

  区域营销的理论基础源于场所品牌推广(Dinnie, 2003; Govers & Go, 2009)、场所体验(Zbuchea, 2014)、场所身份(Zbuchea, 2014)、目的地营销(Hailin et al., 2011)和投资者(Zbuchea, 2014)这些概念。 所有这些政策都适用于具有市场价值的天然资源或者本质的地区的经济增长和发展。根据Zbuchea（2014）所述,为了保证这些项目和政策的有效性,区域营销的模型必须考虑以下关键阶段：确定关键伙伴,设计关键活动,识别资源,提出价值主张,规划客户关系,区分市场,建立沟通渠道和设计预算。长期来看,所有这些都将为该地区带来吸引力和独特性。这也会为与该地区相关的各方带来重要性和价值。

  因此,我们可以将区域营销定义为一种可以用来发展特定区域的有效工具。这些区域定位于通过强化当地居民和组织的活动,货物和服务来促进当地价值和文化的发展。马里布尔发展局（2011）认为区域营销是在一定背景下的战略规划。一个区域必须要先有一种具有潜在价值的重要的资源才能开始营销活动。这就是为什么欧盟（1999）认为"边远地区可持续发展的关键是发展一种独立的视角并发现当地的潜能"。托斯卡纳地区遍地都是世界遗产,在这里取得成功是理所当然的事情。

结论

虽然存在一定缺陷,托斯卡纳模式完全可以被引入到其他地区。但它也是一个尚待改善的机制,因为其中尚有当地官僚的指令性而非绩效性的做法,这就导致了国际投资流向其他旅游目的地。因此,对于古代城镇,我们不能像用人般"完全发挥其潜能",这么做本来可能会在应对2008年危机的时候发挥重要作用。

Regeneration of the Historical Villages of Tuscany, Through Conservation & Tourism Development Strategies

Alessandro Melis, Olufunto Ijatuyi

Introduction

The conservation of existing buildings is now motivated by ethical and environmental reasons, beyond the interests of heritage preservation and social identity needs. It is a global priority to minimize the exploitation of environmental resources, including the soil (Foley, 2011). The conservation and the "zero" volume strategies (Hermelink, 2013) must therefore convey into a holistic approach within high density territories such as Tuscany. This region, located in the north-central part of Italy, and known for the highest concentration of UNESCO sites in the world (http://whc.unesco.org/en/list/), represents the ideal territory for the application of advanced sustainable conservation policies both at the urban and the architecture scale. The shift of population to the suburbs, and the progressive abandonment of rural villages, that are extremely valuable for their historical architecture, has led to the touristic regeneration of the rural building stock. Tuscany has become a model of sustainability, and, at the same time, a brand of touristic success, particularly in the Anglo-Saxon countries (UK, USA, Australia and New Zealand), and in central and northern Europe. The conservation designed by tourism, has become the means through which local governments have agreed with the stakeholders the accomplishment of new infrastructures and facilities and have promoted new employment.

Background

The CO_2 emissions depend mostly (30-40% of the total - source) on the buildings, on the way they are designed, constructed, operated and demolished (Ürge-Vorsatz, 2007; Mata, Kalagasidis, and Johnsson, 2013; Ibn-Mohammed, 2013; Ezema et al., 2015). As the world population is projected to grow over 20% by 2050, the upsurge in global warming, which is a result of greenhouse effect, will lead to a 20% lack of soil for food production during the same period. All this can be attributed to sea level rise, desertification and feedback phenomena. Thus, the European Union - the vanguard of policies on sustainability since the eighties of the twentieth century - is promoting both zero-volume and zero-energy policies (Hermelink, 2013). In territories, such as Tuscany in Italy, with a high historical significance, high presence of UNESCO sites, and the highest concentration of UNESCO sites in the world, the need to operate through the conservation of the existing building stock is an obvious choice and gives a advantage of a good economic practice.

The strategic model is quite simple: the historic villages of Tuscany that are largely abandoned by the local population, are transformed into villages for international tourism. The success of the operation is mainly due to the maintenance of the local identity (highly appealing to foreigners), the high comfort of the residences, and the integration with new facilities related to the touristic activities, but accessible to the local population as well. Indeed, it is a "win-win" strategy: the mentioned facilities are negotiation tools used by the local councils and private stakeholders. Sports facilities, conference rooms, libraries, retail activities all became opportunity for urban regeneration and new employment and, at the same time, elements of attraction beyond the tourist appeal.

Case studies

Four projects in particular, among others designed by Heliopolis 21, can be considered as case studies as regards the strategies mentioned above.

The Aulla project condenses some of the regeneration strategies on a regional scale, such as the organic and compact urban development, instead of a low density model based on single houses, spread in the territory, the integration, within the existing urban fabric, of commercial and sport facilities, that are shared by both tourists and locals, the assignment to the ancient and eventual new villages of a specific vocation, in which production and tourism are integrated (artisan artefacts, food and wine, leisure, health etc.).

The new sports center in Aulla, financed by the developer, is meant to be the central element of the project. It is designed for collective use (tourists, locals and national sport teams). It comprises a sports hall, a spa, and various playgrounds that have become the collector between the designed space and the surrounding landscape for outdoors activities (trekking, horse riding etc.).

The golf course, with its annexes (club house, training center etc.), is part of the project both in Aulla and Castelfalfi. Designed, with the use of new technologies to minimize the water consume, it becomes the catalyst around which it is possible to develop tourist and collective activities available to the community. Castelfalfi project initially included two hotels, the convention center, an acquatic center and various facilities. The commercial and handcraft activities and the restaurants, will be fully integrated into the existing structure of the village surrounding the ancient castle.

The intervention of Nocchi focuses on the conservation of the eighteenth-century building complex annexed to the ancient Villa Grazian. The project aims to transform into an old mill building for the production of olive oil. Its warehouse will be used for the storage of agricultural products, and the farmhouse. The outdoor spaces are shared to keep the original unity of the complex intact. The face of the building has been restored in agreement with the philological conservation theories, approached through careful historical researches (literature review and archives) and through the analysis of original materials and colors.

As for Nocchi in Libbiano, its interiors are inspired by contemporary design. The conservation of its materials and the original structures of the "shell" are legible, while modern materials and technological devices aim to give a positive response to the contemporary needs of the users. Due to the high level of seismicity of the Italian territory, an important part of works was dedicated structural retrofitting. On the sides of the foundations, curbs, which are connected to each other, have been placed in reinforced concrete. Even the height of the wall was made of reinforced concrete rings around the perimeter at the prompting of the roof. Welded steel mesh in the floors, cast in situ concrete beams along the perimeter of the roofs, and vertical reinforced concrete columns within the brick and stone walls, contribute to anti-seismic performance of the building. Steal or concrete portals also constitute the reinforcement for the new openings. Furthermore, the energy performance and the thermal and acoustic comfort of the building have been calculated and examined through appropriate simulations that led to the insertion of new layers of insulation and the elimination of thermal bridges within the walls.

Regeneration of ancient villages

The necessity to regenerate cities can be traced to the relapse of industries in Europe after the Second World War

(McDonald, Malys, and Maliene, 2009). More-developed areas and urban centres, on account of their economic activities and other reasons, entice a lot of people from the villages, leading to condemnation of a big part of such villages and peripheral territories to desertion and desertification (Dinis, 2004). Thus, because attention is focused largely on urban development with its attendant challenges, issues relating to the regeneration of ancient villages have not really enjoyed much focus from scholars. Galland, et al. (2016), in a recent UNESCO publication, stressed that even though there is global interest and mass media attention and investment aiming at the nomination of heritage sites in different places, little is recognised about the conservation and management efforts regarding World Heritage. Literature about cities cannot be complete without relevant reference to villages; this is because cities are actually offshoot of villages. Therefore, since some ancient villages have become urban areas and have been affected by urbanization, the regeneration of these villages becomes paramount. Going by the observation of Li, et al. (2014), though there are no apparent borders among them, four different stages (from the 1950s to the 1980s) explain where the concept of urban regeneration developed from. These concepts are urban reconstruction (1950s), urban revitalization (1960s), urban renewal (1970s), and urban redevelopment (1980s). These terms all point to the same goal of regeneration and they are expressed in the villages of Tuscany.

Territorial marketing

The theoretical background of territorial marketing is traceable to the concepts of place branding (Dinnie, 2003; Govers & Go, 2009), place experience (Zbuchea, 2014), place identity (Zbuchea, 2014), destination marketing (Hailin et al., 2011) and stakeholders (Zbuchea, 2014). All these are strategies targeted at the economic growth and development of certain regions that possess marketable natural endowment or essence. According to Zbuchea (2014), to guarantee the effectiveness of these programmes and strategies, a model of territorial marketing has to consider the following key stages: "defining key partners, designing key activities, identifying resources, suggest the value proposition, plan customer relationships, segment the market, set communication channels and budget design." In the long run, all these would provide the power of attraction and uniqueness or identity to such territory. It will also engender importance and value to all those associated with the territory.

Thus, territorial marketing can be defined as a valuable tool that is employed to address or manage territories aimed at the development of their identifiable and specific values through the enhancement of activities, goods, and services, of people and organizations working there. According to Maribor Development Agency (2011), territorial marketing is known as a set of marketing actions carried out by a group of supporting organisations within the context of the procedure of strategic planning. Before a territory can be marketed, there must exist, a commodity of importance, which is potentially valuable. This is why the European Commission (1999) averred that the "key to the sustainable development of rural regions lies in the development of an independent perspective and the discovery of indigenous potential." No wonder the villages in Tuscany are places to be reckoned with, on account of the presence of significant world heritage in them.

Conclusions

The Tuscan model is certainly exportable in other contexts, albeit with limitations. Also it is a model not yet well-oiled due to the traditionally conservative role of local superintendents and the councils bureaucracy based on prescriptive rather than performance approaches, slants the interest of international stakeholders to the other tourist sites. You can not 'therefore fully exploit the potential' in terms of employemnt from the regeneration of ancient villages, which instead could have a decisive role in overcoming the subsequent economic crisis to 2008.

Bibliography

Dinnie, K. (2004). Place branding: overview of an emerging literature. Place branding, 1(1), 106-110.

Dinis, A. (2004). Territorial marketing: a useful tool for competitiveness of rural and peripheral areas. A paper presented at the 44th Congress of the European Regional Science Association: "Regions and Fiscal Federalism", 25th - 29th August 2004, Porto, Portugal.

Ezema, I. C., Olotuah, A. O., & Fagbenle, O. I. (2015). Estimating Embodied Energy in Residential Buildings in a Nigerian Context. International Journal of Applied Engineering Research, 10(24), 44140-44149.

Foley, J. A., Ramankutty, N., Brauman, K. A., Cassidy, E. S., Gerber, J. S., Johnston, M., ... & Balzer, C. (2011). Solutions for a cultivated planet. Nature, 478(7369), 337-342.

Galland, P., Lisitzin, K., Oudaille-Diethardt, A., Young, C. (2016). World heritage in Europe today. France, Paris: United Nations Educational, Scientific and Cultural Organization (UNESCO)

Govers, R., & Go, F. (2009). Place branding: Glocal, virtual and physical identities, constructed, imagined and experienced. Palgrave Macmillan.

Hermelink, A., Schimschar, S., Boermans, T., Pagliano, L., Zangheri, P., Armani, R., ... & Musall, E. (2013). Towards nearly zero-energy buildings.Definition of common principles under the EPBD. Final Report. Ecofys by order of the European Commission.

Ibn-Mohammed, T., Greenough, R., Taylor, S., Ozawa-Meida, L., & Acquaye, A. (2013). Operational vs. embodied emissions in buildings—A review of current trends. Energy and Buildings, 66, 232-245.

Li, L. H., Lin, J., Li, X., & Wu, F. (2014). Redevelopment of urban village in China–A step towards an effective urban policy? A case study of Liede village in Guangzhou. Habitat International, 43, 299-308.

Maribor Development Agency (2011). The ADC Territorial Marketing Strategy. Retrieved from www.southeast-europe.net/document.cmt?id=148

Mata, É., Kalagasidis, A. S., & Johnsson, F. (2013). A modelling strategy for energy, carbon, and cost assessments of building stocks. Energy and Buildings, 56, 100-108.

McDonald, S., Malys, N., & Maliene, V. (2009). Urban regeneration for sustainable communities: A case study. Technological and Economic Development of Economy, 15(1), 49-59.

Qu, H., Kim, L. H., & Im, H. H. (2011). A model of destination branding: Integrating the concepts of the branding and destination image. Tourism Management, 32(3), 465-476.

Ürge-Vorsatz, D., Danny Harvey, L. D., Mirasgedis, S., & Levine, M. D. (2007). Mitigating CO_2 emissions from energy use in the world's buildings. Building Research & Information, 35(4), 379-398.

Zbuchea, A. (2014). Territorial marketing based on cultural heritage. Management and Marketing, 12(2).

孙以栋
Sun Yidong

浙江工业大学艺术学院高级工程师、硕士生导师，近十年来从事历史建筑与街区、城镇化人居环境设计实践工作。完成和在研国家级、省部级以上课题10余项，发表相关论文20篇（其中核心类论文4篇），指导研究生完成古镇村落论文20余篇。考察路程30余万公里。主要研究方向：传统人居文化研究，文化创意产业研究。

2010　浙江工业大学和山堂艺术设计与文化遗产保护研究中心主任
　　　同济大学建筑与城市规划学院、国家历史文化名城研究中心访问学者
2015　浙江工业大学小城镇城市化协同创新中心社会服务部
　　　日本千叶大学访问学者

College of Art, Zhejiang University of Technology ,Senior Engineer, Academic Supervisor for Master Students

Sun has engaged in design and practices of historic buildings and blocks as well as human settlements environment in urbanization in recent ten years, having carried out over 10 national, provincial or ministerial projects, publicated 20 papers in this field (including 4 core paper), guided master students in finishing over 20 papers concerning ancient towns and villages and covered a distance of over 300,000 km of research trip.

Research Interests: traditional human settlements culture, cultural and creative industries

2010　Director at Heshantang Art Design and Cultural Heritage Protection Research Center, Zhejiang University of Technology
　　　Visiting scholar at College of Architecture and Urban Planning, Tongji University, National Historical and Cultural City Research Center
2015　Social Services Department, Collaborative Innovation Center of Small Town Urbanization
　　　Visiting scholar at Chiba University, Japan

困境与期望
——中国传统村落文化保护与活化路径的实践探索

孙以栋、郑渊晗

【摘要】中国传统村落是世界文化多样性的景观之一，承载着令世界景仰的五千年农耕文明。伴随着工业化、新型城镇化的发展，中国传统村落正迅速失去原有的人情风貌，如不加以合理的保护发展，将会变为文明乞儿和文化乞丐。传统村落文化的保护与活化是一项庞杂且繁难的社会文化经济综合工程。结合相关调研实践，旨在探索中国传统村落文化保护的科学机制、因地制宜的发展模式，在保护中求发展。

[Abstract] Chinese traditional villages are one of landscapes with cultural diversity in the world, which are bearing the five thousand years of agricultural civilization admired by the world. With the development of industrialization and new-type urbanization, Chinese traditional villages are losing their original local customs and features. If they are not protected or developed reasonably, we will become beggars of culture. The protection and activation of traditional village culture is a complex and difficult comprehensive project of society, culture and economy. The purpose of this article is to explore scientific mechanism for the protection of Chinese traditional village culture, development patterns to suit to local conditions, and the development in the protection combined with relevant survey, research and practice.

　　中国传统村落是乡愁最重要的载体，是物质与非物质文化遗产的综合呈现地。中国传统村落是一个群体、社区、宗族、民族的集体记忆和智慧结晶。中国传统村落原住民的衣食住行、宗教信仰、婚丧嫁娶等传统民俗文化是民族文化的源泉与核心。

　　截至2014年，住建部、文化部、国家文物局等部门共同评审出三批共2555个村庄列入中国传统村落名单。在未来的三年，中央财政将投入超过百亿资金来帮助全国3000个传统村落编制保护发展规划、探索修复、保护文化遗产，全方位推动传统村落的保护。

一、中国传统村落的现状

　　中国传统村落从择地建村伊始，直到村落的形成、演变及其发展历程，都在不同程度上受到了社会、经济、自然、宗教、文化等多方面因素的影响，不同地域在发展过程中所处的整体环境存在着差异，因此形成了各具特色的传统村落格局和地域文化。然而，经济结构的快速转型升级，给传统村落带来了巨大的冲击，中国自然村总数由2002年的363万个，骤减到2014年的252万个，短短十年间锐减了近110万个自然村，村落正在快速地消亡。一方面，由于社会产业结构的转变，城镇化的迅速发展，大量的农村人口涌向城市，导致中国传统村落空心化、老龄化。另一方面，在全球经济一体化的时代背景下，本土文化受到巨大冲击，千城一面的城市建设已严重威胁着传统村落的保护和发展。

　　中国传统村落是中国村落历史文化的"活化石"[1]。但是，工业文明正在急速地破坏农耕文明，传统村落中的农耕文化、传统习俗、生产生活方式、古民居正在消失，承载着中国五千年文明的传统村落在被毁灭的过程中，中国传统村落所独具的传统文化非常脆弱且不可再生—保护中国传统村落正成为社会的普遍共识。

二、中国传统村落文化保护发展问题现状

　　新型城镇化发展过程中，村落合并直接致使村落传统风貌消失，这也是村落快速消亡的最关键原因[2]。中

国的传统村落正面临着大规模的破坏和整体空间的更新，在更新的过程中失去了自身的地域文化特征，走着千村一面的城市化发展模式。新型城镇化以来，中国传统村落保护已得到了社会各界的普遍重视，习近平总书记提出"留得住青山绿水，记得住乡愁"。2014年，住建部、文化部、国家文物局、财政部颁布的《关于切实加强中国传统村落保护的指导意见》[3]，中央财政将统筹国家重点文物保护、美丽乡村建设、非物质文化遗产保护等专项资金，来全力支持中国传统村落的保护与发展。

但是由于社会各界对于中国传统村落认识的差距，中国传统村落自然生态和文化遗产的破坏仍然在继续。

1）地方政府是传统村落文化保护与活化的主要执行者。但政府职能部门对中国传统村落历史文化了解的深度不够，各部门之间缺乏协调，各自为政，缺乏系统、科学的规划，建设方式简单粗糙，只重外观形态建设忽略文化内涵保护。同时没有构建合理的产业结构，传统村落保护与活化缺少可持续发展的保障。

2）原住民是中国传统村落的主人，但因人口结构文化素养的局限，在追求现代化生活与保护传统居住文化的矛盾中缺乏有效的解决手段。一方面，随着社会经济的迅速发展，传统村落的更新是一种必然的趋势，原住民也拥有享受现代生活的权利，这必然与原有的传统生活方式产生矛盾。另一方面，由于原住民普遍文化程度不高，对当地村落传统文化资源缺乏有效的认知，缺少有效合理的更新手段，因此，会经常有不自觉地破坏古建筑等不利村落文化资源保护的行为。

3）传统村落商业化运行中不同程度加速了传统村落文化破坏。传统村落的发展需要经济的支撑，但在商业开发过程中，多数投资者一味地追求商业利益的最大化，缺少对地域文化的尊重，从而破坏了原有传统村落的人居环境生态，因此在传统村落发展过程中出现了逐渐同质化的古村镇人居风貌，原住民的生活方式变成舞台表演。

4）传统村落一直是艺术家设计师的创作灵感源泉之地，随着传统村落保护发展关注度不断提升，越来越多艺术家设计师投身到传统村落保护与发展的建设领域，本是令人鼓舞和欣慰之举，但却以文化的优越感俯视乡村，大肆抒发自我情怀，无视村落文化的地域性和多样性。

传统村落是五千年农耕文明进程的瑰宝，历史人文价值极其宝贵，在村落活化过程中要始终将原住民的生活起居和利益摆在第一位，将传统文化真正地融入到原住民生活的方方面面，而不是一味地为了迎合体验者而进行的传统文化展示，这样才能够实现传统文化的可持续传承。对于体验者而言，只有真正地参与到传

图1（左）千村一面，图2（中）表演人生，图3（右）有人无居

图4（左）风俗活动，图5（中）傈僳族节日，图6（右）侗族篝火晚会

统文化体验中去，才能够更好地了解和感受当地的特色文化和传统民俗。

三、中国传统村落活化的多方位实践探索

传统村落凝结着具有五千年悠久历史的传统文化，保持着最原始的中华民族文化形态，对于传统村落的保护要因地制宜、循序渐进地展开，进行多方位的探索。通过传统村落的活化，使传统文化能够得以延续，实现传统文化与现代文明的历史对话和共同发展。在活化过程中可以将体验经济结合进去，为体验者营造出独特的感官体验及思维认同。体验经济从生活本身出发，与原住民的生活又紧密结合，既可以很好地保证原住民的正常生产生活又可以保证原住民的经济来源，形成良好的保护和活化可持续发展。

1）深入学习国家政策，协同多方组织机构，加强传统文化教育。首先，要深入学习理解国家政策，同时通过各领域研讨会、国际论坛整合政府部门、科研机构进行传统村落保护与活化理论实践研究，为传统村落保护与活化的实践探索提供理论依据和智力支持。其次，传统文化的复兴必须大众的教育先行，民族的传统文化要想得到传承、发展和活化，最终还是要依靠民族自己的思想觉悟和实际行动[4]。一方面，充分运用多媒体的舆论覆盖面，加大传统村落保护重要性的宣传，加强传统文化教育，可以提高对传统文化的认知，重建文化自信。另一方面，通过传统文化教育组建多专业、多学科的团队，深入挖掘地方文化，有利于传统村落的保护。

2）活化实践要与地方政府紧密合作，积极推动多专业、多部门之间的联动。积极推进规划先行，分类指导的建设要求，在实践过程中不断推动社会福利、文化设施、医疗卫生产业、建筑保护的发展，在优化产业结构、发展地方经济和挖掘地方文化等方面实现多轨合一。传统人居文化环境建设既是物质建设，也是文化建设[5]，综合各政府职能部门之间的发展要求，在传统村落规划发展上要做到因地制宜、分类指导，从区域角度和地方角度上统筹传统村落文化的保护与活化，注重保护传统文化和周边自然生态环境，加快基础设施和公共设施的建设。

3）原住民是传统村落的核心，他们是传统村落文化保护与活化的关键。深入了解原住民的现实需求，同原住民深入交流，了解原住民的生活现状，带动原住民共同推动村落文化的发展和传承，培养传统手工艺

图7（左）古镇童年，图8（右）老何所依

技术，实现习俗传承，发动原住民运用原有的传统技术进行村落建设，在保留真实性的基础上进行保护与活化。传统村落的文化包括物质和非物质，物质的屋舍、院落、古道等固然重要，而发生于这些物质的存在中的历史、习俗、礼仪等非物质则更加重要，并且这些根植于传统村落中的文化基因随着现代生活的变化可以健康地生长，一个"生活着的村落"能够更好地保护传统村落的文化。

4）处理好传统村落商业开发中对传统村落文化保护与活化的双面性作用。在传统村落的商业开发中传播保护与破坏建设共存，生搬硬造的"假古董"无法引起社会的共鸣，没有对传统历史文化、对原住民生活的尊重，也就不能构成乡愁[6]。传统村落的保护与活化必须回归传统文化，结合社会经济发展，重视文化生态和文化资源保护，变静态保护为动态营造[7]。传统村落的保护不能是静态保护，村落"博物馆"并不能真正起到文化保护与传承的作用，而是应该进行动态地保护，挖掘出一条原生态文化和体验性文化的传承模式。文震亨在《长物志》中指出"若徒侈土木，尚丹垩，真同桎梏樊槛而已"[8]，因此在商业开发中要注重村落自身的传统文化，不应注重大兴土木，崇尚求新求洋，而应结合现有现状，以体验经济为主要开发模式，让旅游观光者参与到传统文化体验中去。

5）建立完善的传统村落文化保护与活化管理体系。一方面，加快对传统村落各类文化遗产现状的立档调研工作，从文化生态保护、自然生态保护的角度探索古村落人居环境可持续发展的要素[9]。对于重要的文化传承人和手工艺技能要采取针对性的保护措施，对传统村落在历史时空信息的留存进行综合研究，建立完善的传统村落保护项目资料库。另一方面，要引导原住民发挥主体作用，参与监督管理传统村落的保护发展规划和传统村落活化建设，调动原住民参加村落文化保护与活化的主动性。

传统村落文化的保护与活化是一项庞杂且繁难的社会文化经济综合工程[10]。建立科学的保护机制，完善的活化措施，探索因地制宜的发展模式，提升传统文化的保护意识，真正做到全民自觉参与。坚持真实性、完整性、多样性，在传统村落文化保护与活化中求发展，在传统村落发展中促保护与活化。传统村落文化的保护与活化为中国传统文化延续千年的脉络，为传统文化招魂。将体验经济的发展潜力发掘出来，更有利于传承传统文化和历史记忆的保存，也有助于早日实现"望得见山、看得见水、记得住乡愁"的现实要求，为传统村落文化的保护与活化探索出一条可持续的实践路径。

参考文献

[1] 刘馨秋,王思明.中国传统村落保护的困境与出路[J]. 中国农史, 2015 (4): 99-110.

[2] 冯骥才.传统村落的困境与出路——兼谈传统村落是另一类文化遗产[J].传统村落,2013 (1): 7-12.

[3] 彭泺.四部门联合下发指导意见加强中国传统村落保护[J]. 建筑设计管理, 2014 (5): 41.

[4] 单霁翔.保护民族文化遗产维护文化多样性[J]. 中国民族, 2014 (2): 28-33.

[5] 吴良镛.人居环境与审美文化——2012年中国建筑学会年会主旨报告[J]. 建筑学报, 2012 (12): 2-6.

[6] 阮仪三.留住乡愁——新常态下文化原乡的回归[J]. 城乡建设, 2015 (9): 9.

[7] (明)文震亨著,陈植校注.长物志校注[M].南京:江苏科学技术出版社, 1984: 18.

[8] 潘鲁生.传统村落的活化与发展——写在中国民间文化抢救工程巡礼之际[J].设计艺术, 2015 (3): 61-64.

[9] 孙以栋,王添翼.协同创新机制下古村落人居环境保护路径研究[J]. 建筑与文化,2015 (7): 135-136.

[10] 刘晓路."中国古村落保护"国际高峰论坛在浙江举行[J].民间文化论坛,2006 (4): 75.

诺伯特·施密茨
Norbert M. Schmitz

德国基尔穆提休斯艺术学院美术学教授,媒体艺术历史学家,任课于德国伍珀塔尔、波鸿、奥地利林茨、瑞士苏黎世、萨尔茨堡等多所大学和艺术学院。在美国芝加哥、伯克利、纽约、亚特兰大、明尼阿波利斯、巴黎、苏黎世、伯尔尼、萨尔茨堡、新德里、拉合尔、加德满都、首尔多所大学和艺术院校教授国际课程。研究方向主要为艺术与电影之间的图像学和中间性,新媒体与旧媒体的图像学,艺术体系和全球艺术的理论研究。

学术代表作
《当代符号的艺术与科学》,魏玛 1994
《运动作为符号形式》发表在《图像:媒体科学界的观点》,马尔堡 2000
《交互设计——设计作为互动形式:论设计作为现代交际的符号形式的紧迫性》发表在《交互设计》,Birkhaeuser 2005

Dr. phil., Professor for Ästhetics, at the Muthesius Academy of Fine Arts and Design, Kiel. Art- and Mediahistorian, Teaching at Universties and Art-Academys in Wuppertal, Bochum, Linz, Zürich und Salzburg. Working about Iconology and Intermediality between art and film, Iconology of old and New Media, Medien, Discourse theory of the Art- and Design system.

Publications:
Kunst und Wissenschaft im Zeichen der Moderne, Weimar 1994;
Bewegung als symbolische Form, in: Über Bilder Sprechen. Positionen und Perspektiven der Medienwissenschaften, Marburg 2000, S. 95-135;
Interactivedesign – Design as Interaktion. On the emergency of Design as an symbolic Form of the modern Communication, in: Interactive Design, Birkhäuser 2005;
Hopper`s Modernity, in: Western Motel. Edward Hopper and Contenmpoary Art, Cat. Kunsthalle Wien, Nürnberg 2008, P. 240-259;

以真实性为教条
——呼吁以实用的方法保护历史建筑

诺伯特·施密茨

说到保护历史建筑或者保护世界遗产，以我这个德国人的视角来看，中国目前的情况还是非常戏剧性的。我这么说就好像我对中国独特的文化遗产的现状并不那么担心，而更关心遍布城乡的传统历史建筑和构架，它们因为中国的大规模现代化建设而随时面临消失。

总的来说，德国的历史建筑的保护仍然主要包括历史建筑的存续和经济以及社会政治的限制。

我认为思考历史建筑的保护有助于我们更好地理解目前迫在眉睫的城区重建和农村城镇化与历史遗产保存之间的关系，后者是当地文化的重要组成部分。我希望建筑师和保护主义者都不会在此次会议上用这种有关保护的文化科学前提条件的观点参与讨论。

但是在理想状态下，我们会看到两种截然不同的观点。

一种观点认为人们应该完全保留历史建筑的旧貌，新旧建筑区别鲜明。这就意味着对历史建筑的任何形式的翻新都会造成问题。除了原貌以外，其他修饰都是虚假的。

另一种观点认为，历史建筑是当今社会的一部分，和诸多现代元素结合在一起。为了适应当今社会的需求，这些建筑通过经济资源得到重新修缮和阐释。双方的观点都很有道理。

在中国说到遭破坏的历史建筑群，人们自然会想到被拆毁的胡同。在我看来更为实际做法是减少可能无法调和的矛盾，并且扩大文化遗产的范围，不仅限于散落于杂乱的时髦建筑中的几件博物馆展品或者突兀于迪斯尼乐园中的一堆形体。

但是，这就意味着否定当代的激进历史建筑保护主义，该主义遵从完整保存建筑的历史风貌。对于现在的当地居民来说，这种理念不切实际，开支过高且格格不入。最坚定的保护主义者们现在的想法是如何形成的？ 这源于对完全的原真性和"真实的建筑"的求索，它们是现代性的美学标准，也是现代设计和建筑理论的特点。

历史建筑的保护最早源于浪漫主义时期和历史主义，兴起于19世纪早期并在其间得到发展。它不仅确定了对历史传统的保护，而且还形成了当代建筑和都市主义。

我们拿美国和欧洲的城市举个例子。巴黎、伦敦和纽约都是当今都市主义的典范，但其实这都源于大力保护历史建筑。这些城市的历史建筑群来自各个年代，有些甚至可以追溯到浪漫主义和哥特时代。人们应该意识到19世纪以及这些历史建筑的重要意义，它们成就了这些城市的独特魅力。

最坚定的保护主义者们现在的想法是如何形成的？ 这源自于对完全的原真性和"真实的建筑"的求索，它们是现代性的美学标准，也是现代设计和建筑理论的特点。

历史建筑的保护最早源于浪漫主义时期和历史主义，兴起于19世纪早期并在期间得到发展。它不仅确定了对历史传统的保护，而且还形成了当代建

图1 杭州西湖湖畔的重建房屋

图2 印度加尔各答的殖民地区海滨　　　　　　　　　　　　图3 伍珀塔尔市的保护街区

筑和都市主义。

但是欧洲以外的情况也是这样。欧洲城市文化的帝国主义扩张在全球各地都得到了不同的展现，不论这些地区是否曾经直接或者间接受到欧洲殖民的影响。各地的人们吸收了当时的现代技术和都市规划概念，并在改造城市的过程中为其赋予当地文化特征。但这些都不是原创或者真实的。当地的历史建筑最多就成为一种正式的点缀，并且经常成为蔚为壮观的城市地标建筑。在拉合尔和德里，英国人重修了莫卧儿王朝时期的建筑，并在其周边兴建了受此启发设计的伪印度殖民风格建筑。

这样看来，我们只需要讨论下此类五座最有名的欧洲和美国的大都市：

这五座城市代表着纷繁各异的殖民压迫历史，但与此同时，它们又有着鲜明的国家特色并被视为国家遗产，重要性不亚于佛罗伦萨和阿格拉时期。我稍后会讨论这个问题。

现今的建筑或者设计是如何往这个方向发展的？历史主义的艺术是一个自成一体的怪物。对于现代主义者来说，对于独特而演进的艺术的大量复制更多是出于经济考虑，而不是创造"真正的建筑"。

如前述的那些城市所代表的19世纪的城市概念被认为是"不真实的"，因为它们使用了最新的科技来呈现代表过去的形体。但是，对于光晕艺术的执迷本身就是非常现代的。具有浪漫主义倾向的前卫派理论家瓦尔特·本雅明首先阐述了历史的光晕艺术，这源自《机械复制时代的艺术作品》的现实情况而导致的他对艺术的旧定义的误判。人们可以总结道："在机械复制艺术时代凋亡的不论何物都是它自己的光晕。"[1]

不同于以历史主义的形体语言利用现代科技任意打造各种形态，在工业生产中完全忠实于材料的做法赢得了推崇和艳羡。材料凸显了"真相"和"伪造"之间的区别。[2]

柯布西耶在印度的设计同样缺少对于传统美学的思考。这并不是因为他没有意识到这种传统，而是因为他不想在历史和绝对的现代性之间做出平庸的妥协。

本雅明简单地总结出了他所处时代的现代艺术家和建筑家的创作态度。可重现性似乎使得传统工匠技巧变得多余且失去了相应的地位。因为失去了来自教堂和政府的委托作品，富有自主性的现代艺术家处境艰难。工业化生产艺术品日益普及，艺术品的好坏不再取决于传统艺术家的水平，其价值就受到了质疑。

正是此原因，"真正原创"成为了与著作权相关联的指导性原则。[3] 这种概念映射在了历史上。最终，人们开始禁止使用现代技术来重建历史遗迹。取而代之的是一种意识形态，认为历史遗迹消失殆尽，因为我们无法以"绝对形态"进行保留。

或者我们可以更富争议性地说：如果不能完全恢复历史建筑的旧貌，就拆掉它！

因此，我们不应该把"原真性"的错觉和历史混淆起来。历史主义的看法是如此地不同！

历史主义者随意解读其所处时代的剧变，并将其视为一种机会以及对都市建筑的一种创造性的形体影响，以简单地将这种形体词汇和当时最先进的技艺联系起来。他发展出了历史建筑保护所需的科学先决条件，而不会因为自身的现代性而受到尼采所谓的古物研究史学观点限制。[4]如今，我们将出自19世纪的明显"非真实"建筑视为真实的历史遗产，与此同时，

图4 重新修缮的颐和园，北京

我们也无法将历史主义作为一种融合到普通建筑中的先决条件，开放地讨论历史传统。这似乎值得商榷。

表面伪装于是变成了真相。这就说明了技术几乎可以实现任何事情。如今人们已经不再关注创造"纯正的建筑"，而是构建一种可以表述我们精神和物质需要的语言。[5]

文明进程中的都市性正是源自一种想象的空间。从这种意义上来说，以巴黎和纽约为典范的19世纪的城市概念是无法被超越的。这对于当今的社会和中国来说意味着什么呢？

只有在技术和经济发展迅速的今天，我们才能普及这种城市概念，而它曾经只限于富庶的资本主义地区且最终周边环绕贫民区。不论其历史地位，这种较之20世纪20年代的乌托邦更加实际的观念在我看来是最为世故的，如同Bazon Brock曾经评价的："西方社会这两种截然相反的观点之间已经出现了对峙，一方面，可悲的包豪斯功能主义和乌托邦将自己与之联系起来，而另一方面又是百货商店文化的随意性。我认为现在的潮流倾向于允许自由使用形体的历史语言，而不是假装我们现在还能创造出和19世纪初一样的经典建筑。"[6]

亚洲的人们破坏历史遗迹，然后用最新或者最便宜的技术和现代材料重新修缮，这经常会在欧洲引来众怒。比如日本的某些寺庙在几十年之后被拆毁然后按照原来的样式重建，以及巴厘岛上使用标准水泥构件扩

图5 河内新建的普通建筑区域

图6 Wuppertal 重建的普通建筑区域

49

图7 杭州西湖湖畔重建的庙宇　　　　　　　　　　图8 上海古镇，遗迹或者重建

建和重建的古老庙宇。同样地，对于修缮和重建来说，法国殖民风格的建筑外形构件也很容易获得，这成就了如今的河内市貌。于是就出现地区文化以及具有地方特色的建筑风格和惊艳的城市景观，这完全不同于遍布全球的现代化摩天大楼建筑群。

我再重申一下：只有将"真实性"视为现代性的一个组成部分，才能明白其中的挑战。以前的艺术家和工匠并不明白本雅明认为已经失去的光晕是什么。我们如果认为这非常重要并需要将其投射到过去的历史上，那我们就要非常小心。我们更加应该认识到历史一直以来都是我们根据当下的利益构建出来的。

因此我们担心的并不是保护"真实的"历史，而是将历史融入到现代生活中现实可行的形体中。在历史遗迹保存这个问题上，我希望从纯唯物的角度对德国唯心主义学说中著名的"真善美"稍作改动，我们应该关注"美"而让哲学家们来讨论"真"的问题。

参考文献

[1] Cf.: Benjamin, Walter: Das Kunstwerk im Zeitalter seiner technischen Reproduzierbarkeit, Frankfurt a.M. 1963/1977, S. 13.

[2] Cf. for example: Loos, Adolf: Ornament und Verbrechen (1908), in: Ders. Sämtliche Schriften 1962, p. 277.

[3] On the construction and ambivalence of the term authenticity in detail: Schmitz, Norbert M.: Der Diskurs über Performance und der Mythos des Authentischen. Eine Kunstform als Übung zivilisatorischer Alltagsästhetik, in: Performance im medialen Wandel, ed. by Petra Maria Meyer, München 2006, pp. 441-462.

[4] Cf. Nietzsche, Friedrich: Vom Nutzen und Nachteil der Historie für das Leben. Unzeitgemäße Betrachtungen II, in: Ders.: Werke. Krit. Collected Edition, edited by Giorgio Colli and Mazzino Montinari, Third Part – 1st Volume, Berlin/New York 1972.

[5] Cf. Robert Venturi, Denise Scott Brown and Steven Izenour: Learning from Las Vegas, Revised Edition. The Forgotten Symbolism of Architectural Form, MIT University Press Cambridge, Mass., 1977.

[6] Bazon Brock: Die Frage nach den Bedürfnissen - am Beispiel des Wohnbedürfnisses, Vortrag auf der Loccumer Tagung, Wohnbedürfnisse der Zukunft' 1975, in: Ders.: Ästhetik als Vermittlung. Arbeitsbiographie eines Generalisten, Köln: DuMont, 1977, p. 419. Translation by Colin Moore Kiel.

Authenticity as Dogma – A Plea for the Pragmatic Conservation of Monuments

Norbert M. Schmitz

When we talk about monument conservation or about saving our world heritage, from a German perspective the situation in China seems to me to be quite dramatic. Saying this upfront as it were, I am less concerned with the unique cultural heritage of Chinese civilisation than with the broad field of traditional historic buildings and structures in urban and rural areas – in the theme of this conference – which are almost inevitably threatened with disappearance due the tremendous dynamism of China's modernisation.

Generally, monument conservation in Germany continues to move between the idea of preserving monuments of the past, and economic and socio-political constraints.

I believe that self-reflection of the ideas of what monument conservation was and is may be helpful here to develop ideas for a position which seems appropriate to negotiate between the urgent need for urban renovation, the modernisation of rural habitats and the conservation of that heritage which is an indispensable part of cultural identity. Neither architect nor conservationist do I hope to enrich this conference with such a view of the cultural-scientific pre-conditions of current discourses on conservation.

However, ideally, two contrary positions can be seen.

On the one hand, those who demand a radical preservation of the original status as a complete document of history, in which there can be no doubt what is old and what is new. This means that any great modernisation or compromise with econo1mic, stylistic or social requirements is problematic. Anything else would be a lie in this respect.

On the other hand, historical monuments are seen as part of a social and aesthetic present, are combined with them and due to the different requirements of everyday life not only rebuilt but re-interpreted using economic resources.

Both positions can lead a strong case for themselves.

In the face of the concrete destruction of historic ensembles–in the case of China one thinks of course of the Hutong Quarter–it appears to me that a pragmatic position is developing in which the probably irresolvable contradiction is at least reduced and cultural heritage is not limited to a few museum objects ending up in an unprincipled desert of modernism in one side or a potpourri of forms à la Disneyworld on the other.

However, this means saying goodbye to radical ideas of modern monument conservation according to the absolute authenticity of historical tradition since it

Fig.1 Restored Building at the West-Lake in Hangzhou

Fig. 2 Colonial Stil in Calcutta, The strand, Foto: Regina Höfer Bonn 2012

Fig. 3 Restored street in Wuppertal, Foto: Stephan Dückers, Wuppertal 2007

is simply impractical, unaffordable and often incompatible with the people's legitimate demands concerning their architectural habitats.

How does the idea which most dedicated conservationists still have in their minds today come about? It is the search for complete authenticity and 'true architecture' as the aesthetic programme of modernity which characterises the notion of modern design and architectural theory.

Monument conservation is firstly an idea of the Romantic Period and of Historicism, in example it arose at the beginning of the 19th century and developed throughout it. It formed not only the assurance of tradition but also contemporary architecture and urbanism.

An example: The European-American city as a much admired reference even of today's urbanism, as can be found in cities such as Paris, London or New York, is anything but the result of radical historic conservation. So rich in tradition such cities are with regard to their core monument inventory up to the Romantic and Gothic period, one should realise how great the proportion of the 19th century and historistic architecture of these townscapes is which characterises their attractive images today.

This is however true outside Europe. The imperialist expansion of European city culture transplanted this type of city throughout the world irrelevant of how each country was directly or indirectly colonised by imperial powers. In doing so, one adapted the then modern technologies and urban planning concepts to give them a local character with typically local or regional ornaments. None of this is however original or authentic. The monuments of the country became at best formal stooges and often impressively conserved into significant city landmarks. Thus, in Lahore and Delhi, the British restored the famous Moghul monuments to surround them with their own Moghul-inspired, pseudo-Indian colonial style.

In this regard, one should simply mention the five perhaps best-known international Euro-American metropoles of this type:

All five examples represent the most diverse forms of dependence and colonial oppression, but at the same time, today they are places of national identification and rightly seen as a national heritage no less than Florence or Agra are. I shall come back to this.

How does today's architecture or design behave towards this? The art of Historicism is its bogeyman sui generis. For the modernists, this infinite reproduction of the unique and the evolved was about mass production as an economic consideration and not about 'real architecture'.

City concepts of the 19th century such as those already mentioned were considered 'inauthentic' because they implemented symbolic forms of the past by means of the latest technology. However, the

Fig. 4 Peking Restored situation in the Summer-Palace
Foto: Regina Höfer Bonn 2011

fascination of the auratic is itself utterly modern. It was the theorist of a romantically inclined avant-garde, Walter Benjamin, who was the first to discover the idea of the auratic art of the past, in his misjudgement of the old concept of art due to the reality of 'The Work of Art in the Age of Mechanical Reproduction'."One can summarise what is noticeable here using the term aura and say: whatever withers in the age of the mechanical reproducibility of artwork is its aura."[1]

An absolute material justice along the lines of industrial production as opposed to the language of form of historicism which realised any form at will using the latest technological resources was met with the highest esteem and admiration. The material highlighted the difference between 'truth' and, 'forgery'. [2]

And accordingly, for India, Corbusier's designs gave little consideration to aesthetic traditions, not because their creator would have had little sense of this tradition but because he did not want to enter into a bland compromise between history and its claim on absolute modernity.

Benjamin summarised simply a basic attitude of modern artists and architects of his time. Reproducibility appeared to make traditional artisan qualities superfluous and took away their legitimacy. This became difficult for the modern autonomous artist since his or her role became problematic due to the disappearance of the classical commissions of state and church. The value of art became questionable since it was no longer guaranteed by traditional criteria of artisan prowess but became increasingly easy to produce industrially. Precisely for this reason, the 'authentic original' became the guiding principle linked to the idea of individual authorship. [3] And this concept was projected onto history. Consequentially, the reconstruction of the past using the means of modern technology became forbidden. Instead, respect for the same turned into ideology with which the same past was eradicated, since its conservation was not possible in 'absolute form'.

Or said more polemically: Demolish a monument rather than not keep it completely original!

Thus, we should not confuse the myth of 'authenticity' with the past. How different things are with Historicism!

Fig. 5 New built common architectural environment in Hanoi, Foto: Regina Höfer Wuppertal 2009

Fig. 6 Restored common architectural environment in Wuppertal, Foto: Stephan Dückers Wuppertal 2011

Historicism responded to the upheavals of his time by freely interpreting the historic inventory and taking it as an opportunity and as the formal creative influence on urban construction in order to simple-mindedly link this vocabulary of form with the most advanced techniques of the times. In doing so, at the same time he developed the scientific preconditions for adequate monument conservation without letting himself be inhibited in his unconditional modernity by an antiquarian historiography in the sense of Nietzsche. [4] Today we experience the apparent 'inauthentic' architecture of the 19th century as authentic heritage and are at the same time unable to implement this free discussion of tradition in the sense of the historicism as a precondition of its integration into ordinary construction. Thus, we lose the ability to keep and update symbolic sites of identification. This seems worthy of consideration.

The lie of disguise is thus its truth. It alone corresponds to the status of technology which simply makes almost anything possible. Today, it is no longer about the construction of 'pure architecture' but about a language in which our symbolic as well as our practical needs are formulated. [5]

Urbanity as the core of civilising processes is precisely the creation of an imaginary space. In this sense, the city concepts of the 19th century culminating in great cities such as Paris and New York cannot be beaten. What does this mean for the present and also for China?

Only the economic and technical resources have exploded to such an extent that today this once limited concept which was that of a well-off capitalist bourgeoisie and one which inevitably carried the dark side of its suburban slums around with it can be created for anyone. Regardless of the historical positions, this conception – comparatively factual considering the utopia of the 1920s – seems to me to be the most worldly-wise as Bazon Brock once suggested: "This confrontation is taking place in the west between two extreme positions; on the one hand, the pathetic functionalism of Bauhaus and the social utopia associated with it; on the other, the arbitrariness of department store culture. I believe that the trend is moving towards allowing historical languages of form to be used again freely – not pretending that one could today still set up in the same classicist manner as

Fig. 7 Rebild Temple area at the West-Lake in Hangzhou

Fig. 8 Original or reconstructed? Oldtown in Shanghai, Foto Norbert M. Sxchmitz Wuppertal 2015

in 1800." [6]

For Europeans it is often irritating to see how original substance destroyed in Asia in order to have it resurrected according to the latest or cheapest technologies using modern materials. I mean here not only famous phenomena such as certain Japanese temples which are demolished after a couple of decades or so and having been identical for centuries resurrected in their old forms. I am thinking here of the standardised components of concrete façades with which historic temples in Bali are extended and sometimes rebuilt. Similarly, for repair and reconstruction, one can easily acquire façade components in the French colonial style, which still characterises Hanoi today. The result is the population's greater identification with their architectural environment and a striking cityscape which differs significantly from the monotony of downtowns all over the globe with their modernist skyscraper architecture.

Once again: The pathos of the 'authentic' itself can only be understood as a construct of modernism; the aura Benjamin believed to be lost was unk nown to the artists and artisans of the past. So we should be careful about projecting it onto the past as an absolute. Rather, we need to understand that history is always something that we construct out of our present interests.

We are thus not concerned with conserving an 'authentic' past but of transforming the tradition into practical, socially satisfying and economically feasible forms of modern life. In a modification of the famous formulation of German Idealism, concerning 'truth, beauty and goodness' as one unit, I would like to say completely materialistically that perhaps in conservation matters we should look for the 'good' but leave the question of 'truth' to the philosophers.

References

[1] Cf.: Benjamin, Walter: Das Kunstwerk im Zeitalter seiner technischen Reproduzierbarkeit, Frankfurt a.M. 1963/1977, S. 13.

[2] Cf. for example: Loos, Adolf: Ornament und Verbrechen (1908), in: Ders. Sämtliche Schriften 1962, p. 277.

[3] On the construction and ambivalence of the term authenticity in detail: Schmitz, Norbert M.: Der Diskurs über Performance und der Mythos des Authentischen. Eine Kunstform als Übung zivilisatorischer Alltagsästhetik, in: Performance im medialen Wandel, ed . by Petra Maria Meyer, München 2006, pp. 441-462.

[4] Cf. Nietzsche, Friedrich: Vom Nutzen und Nachteil der Historie für das Leben. Unzeitgemäße Betrachtungen II, in: Ders.: Werke. Krit. Collected Edition, edited by Giorgio Colli and Mazzino Montinari, Third Part – 1st Volume, Berlin/New York 1972.

[5] Cf. Robert Venturi, Denise Scott Brown and Steven Izenour: Learning from Las Vegas, Revised Edition. The Forgotten Symbolism of Architectural Form , MIT University Press Cambridge, Mass., 1977.

[6] Bazon Brock: Die Frage nach den Bedürfnissen - am Beispiel des Wohnbedürfnisses, Vortrag auf der Loccumer Tagung, Wohnbedürfnisse der Zukunft' 1975, in: Ders.: Ästhetik als Vermittlung. Arbeitsbiographie eines Generalisten, Köln: DuMont, 1977, p. 419. Translation by Colin Moore Kiel.

马利奥·皮萨尼
Mario Pisani

1947年　出生于罗马
1973年　从罗马智德大学建筑专业以优等成绩毕业，毕业论文发表
2000年　开始任职于那不勒斯腓特烈二世大学建筑专业，二等教授
2011年　在那不勒斯和卡普里参加SAVE 历史遗产保护国际论坛，文化遗产保护
2011年　受到中国美术学院邀请，开办二十世纪意大利艺术讲座
2011年　受到中国美术学院邀请，参加认识论和现代艺术交流会议
2011年　在塞尔维亚自治区域Vojodina的当代艺术博物馆开办"2000—2010意大利现代建筑展览"，并举办二十世纪意大利专题讲座
2012年　在多伦多意大利文化研究所举办"2000—2010意大利现代建筑展览"
2012年　在加拿大温哥华举办"2000—2010意大利现代建筑展览"
2012年　罗马圣路卡学院，与Paul Portuguese组织并协调地标景观课程
2014年　指导意大利大学设计竞赛《沙特艾卜哈：旅游之城》
2015年　《Squares of the New Millennium》出版
2015年　在沙特阿拉伯组织"Thi AEEN的修缮和扩张"研讨会

Mario Pisani born in Rome on September8, 1947. Graduated in Rome, Faculty of Architecture, University La Sapienza in 1973 with 110 cum laude and publication of the thesis. Since 2000 he is Professor of Second band at the Faculty of Architecture Luigi Vanvitelli, University of Naples II.

09/10/11 The June 2011 in Naples and Capri participates in the International Forum SAVE Heritage Preservation with the participation of cultural heritage. On 14 and 15 November 2011 invited by the Academy of Fine Arts in Hangzhou, China, participates with teachers from various parts of the world, the seminar on epistemology and contemporary art. And 'course in the published proceedings. From November 7 to 13 in the same place it holds a series of lectures on the Art of the Italian twentieth century. December 8, 2011 at the home of the Museum of Contemporary Art of the autonomous region of Vojvodina, Serbia, Novi Sad presents the exhibition Italy

Now Architecture in Italy from 2000 to 2010 and held a seminar on Italian twentieth century. The exhibition organized by him remains open until December 24. The March 8, 2012 shows the headquarters of the Italian Institute of Culture in Toronto the exhibition he organized Italy Now, Architectures 2000-2010. On March 11, the same exhibition is presented in Vancouver, Canada. From 21 May to 1 June organizes and coordinates with Paul Portuguese at the Accademia di San Luca, Rome, the course and the landscape Mark May 28 participated in the panel discussion on New vistas on the landscape. In 2014 he participates as responsible in the design competition organized between Italian universities: a City for Tourism: Abha, Saudi Arabia. In July 2015 it is published the volume: Squares of the new millennium. In November 2015 in Saudi Arabia organized a design workshop for the restoration and expansion of the city of Thi AEEN.

历史古镇的复兴和农业区域的干预措施

马利欧·皮萨尼

我们不需要不朽的事物。
我们只希望事物不失去他们的全部意义。
——Antoine de Saint-Exupery

你放弃城市中的新居,离开亲戚和朋友,居住到山野之中,沉浸在自然美景之中。
——Leonardo da Vinci, Treatise on Painting, Chapter XX p. 56

充足的劳动力供给使得意大利在20世纪50到70年代之间经历了强劲的经济增长和科技发展。成千上万的居民从意大利南部和北部的贫穷地区迁移到了热那亚、都灵和米兰组成的工业三角区。农民们真正变成了产业工人。由此导致了小型农业中心人口的减少以及大型城市郊区人口的激增。

被他称为经济繁荣的时期造成的这种矛盾延续至今。很多地区的居民背井离乡,大片土地荒无人烟,尤其是在亚平宁山区。这直接导致大片的农田荒废,河道和溪流无人管理,因此出现了山崩等地质灾害。

意大利60%以上的人口居住在市郊,这些区域的经济,社会,人口,城市规划和建筑衰败情况都很不一样。我们发现年轻人和新组成的家庭尤其缺少住房。现有的数据足以让我们深思。意大利国内传承下来的住房有超过2亿套,居民只有大约6千万,这就意味着我们每个人会占有超过3套住房。

很显然,像贝卢斯科尼这样的人肯定拥有远远超过3套的住房。真实的情况是很多意大利人,尤其是住在大城市的居民,连一套住房都没有。很多年轻夫妇和自己的父母住在一起。但是另外一方面,住房库存大,有些小镇因为年久失修可能会完全湮没。

现在似乎有了初步扭转的迹象。老人们开始离开大城市,因为那里物价太高,交通拥挤而且空气质量不好,他们开始回归原籍。而且部分年轻人们由于无法在大城市中找到工作,也回到了农村,带去了新鲜血液。我们想再次讨论两个颇有代表性的地区:圣斯泰法诺迪塞桑约和索罗梅奥。

塞桑约位于阿布鲁佐大区,是一个靠近拉奎拉的小镇。这个小镇可能是从名为Sextantio的罗马定居点分离出来的。关于该区的最早记录是公元760年,宗教社团开发了大片的可耕作农田,人们开始在高海拔地区定居,城镇开始出现并日益扩大。因为这些城镇所处地势较高,也避开了蛮族的侵袭。在新千年的开始之际,从城市归来和留守于此的年轻人为这些古代村落带来了新生。来自瑞士的企业家Daniele Elow Kihlgren 在当地修建一个颇受欢迎的旅店,振兴了当地的旅游业。此举带动了其他投资者的兴趣,他们出资修缮了当地的建筑并开始在当地培育高品质农产品。

古建筑的修缮工作修旧如旧,保留了檐壁,突出部分小径的原貌,历史和传统得以传承。这些房屋都是在中世纪用白色石灰岩修建的,岁月流逝,这些石灰岩的颜色也渐渐变暗。重新开张的小店是修建在山壁中的,是大萨索峰地区最美丽的景色。新的数字科技将小镇和全球连接了起来。当地还提供很多地中海饮食风味的特色菜。美丽的自然风景,静谧的乡村步道和古镇神韵让人流连忘返。Da nobis hodie incantum quotidianum (赐我今日之魔力,天天皆然) 这是一句古代的祷告,说明了咒语如同面包般重要。

索罗梅奥是科尔恰诺下辖的一个小镇,在佩鲁贾附近。这个小镇位于海拔273米的山顶,2001年该地的居民有436人。因为出土过伊特拉斯坎文明的工艺品,所以这个地方的名字源于伊特拉斯坎神灵的名字Lumn。

Brunello Cucinelli SpA公司是山羊绒面料行业的领军企业,他们在山谷中开办了一间工厂。除了针织品展示室以外，因为买家越来越多，山谷中还开办了餐馆和旅店。Cucinelli在索罗梅奥新建并修复了很多文化建筑。他开办了艺术论坛，和平广场，剧院，图书馆和奥勒良学院，并修缮了旧宅和街道，充分体现了John Ruskin的影响。整个地区变成了一个充满灵气的公园。

这位企业家希望恢复并唤醒这些中世纪建筑，这是翻新工作的基础。这样古镇才能成为一个滋养心灵，沿袭文化的地方。随后，他在山谷中的一个废弃工厂中建起了一个针织品工厂。工厂内的工人能够透过大扇的玻璃窗户看到外面醉人的景色。工人在工作的过程中能看到周围的自然风光和自家的住宅，他们的工作就不会那么索然无味了。

那里还有个大约占地0.28平方千米的农业公园，其中有花园，酒庄，橄榄园和其他果园。种植的作物包括小麦，玉米和向日葵。农民们悉心耕种，所产出的蔬果仅供当地消费，送到公司食堂或者当地工人的家中。

第二部分我想讨论下托斯卡纳地区的干预措施。佛罗伦萨市的佩萨河谷圣卡夏诺拥有17247居民，位于Pesa河谷和Greve河谷之间的山顶。这个小镇位于佛罗伦萨以南15千米，锡耶纳以北45千米，Via Cassia从中穿过。它位于盛产葡萄酒和橄榄油的基安帝地区的中心地带。这是一块拥有Officine Graphics Stianti的农业区域，对于当地的经济来说非常重要。

Natalini Associates的建筑师们设计了这个项目。这个区域遍布已经使用过的设备，正好可以用来规划城市的新区。

历史建筑内部都得以修缮，但是外观仍然保持了原貌。这些建筑与停车场和新住宅相连接，紧挨着酒窖。除此之外，当地还修缮了152间临街公寓。这些二楼公寓的尺寸各异，它们的楼下则是店铺。公共步道穿过一楼和庭院，地下则是停车场。各个建筑的一楼都以庭院相间隔，而又都是由公共步道相连接。

默茨河在处于s-Hertogenbosch西北，Engelen 和Bokhovenz之间地方转了个大弯，并且通过运河流向Dieze。 Dieze占地约0.8平方千米，地势平坦，植被遍地。荷兰的Natalini Associates在这里修旧了一个堤坝以保护这片土地。

Sjoerd Scoeters和几位田园设计师设计了Haverleji项目的总体方案。他们用树林和绿色三角地块开辟了这个区域，其中融合了自然和人工因素。他们计划新建一个27洞的大型高尔夫球场，森林中的一块区域，一个自然保护区和大约1000所房屋。1000所房屋中的600所将建造在一个小镇中，其他的房屋将被分到60个称为"城堡"的住宅群中。这些建筑将分布在这个区域的各个节点上。

一个ouNatalini的设计师设计了一个池塘中的"城堡"，用于为其他住宅群作范例。为了起到城堡的保护作用，设计师修改了城堡和防御工事的图纸，他在一条伪内切圆的路径上设计了一扇门，一个弦月窗，一个半圆防御土墙，城墙和角楼。房屋倚防御工事设施而建，外围设有一圈围栏，房屋外侧是池塘，而内部是一个花园庭院。建造房屋所用砖块的颜色也不一样，与水接触的砖块颜色较深，越往上砖块的颜色就越红。大门和角楼是用石头做的，有些墙上用了锌涂层。

勒·柯布西耶说过："建筑是一种艺术事实，是一种感情现象，已经超越了建构的问题。建构是为了支撑，而建筑是为了移动。我们认同并尊重这些规则。当你建立了某种关系时，你就被作品所吸引了。建筑就是一种'关系'，是一种纯粹的精神创作。"看来确实如此。

The Restoration of Historic Towns and Interventions in Agricultural Areas

Mario Pisani

> We do not ask to be immortal.
> We only ask that things do not lose
> All their meaning.
> —Antoine de Saint-Exupery

> You move, or homo, to abandon their new homes in the city and leave them relatives and friends,
> to go to places and rural mountains to valleys, at the natural beauty of the world,
> which, as I consider, with sol sense of seeing enjoy ?
> —Leonardo da Vinci, Treatise on Painting, Chapter XX p. 56

Between the fifties and the seventies of the twentieth century Italy went through a period of strong economic growth and technological development, due to the wide availability of labor. Thousands of people were willing to move from small agricultural centers, where they were born and lived, especially in the south of the Italy and the poorest regions in the north, to the industrial triangle formed by the cities of Genoa, Turin and Milan. There has been a real transfer of peasants agriculture who became industry workers. The consequence was the depopulation of small agricultural centers and the rapid growth of the suburbs of large cities.

Today we are still living the contradictions due to the time, he called the economic boom. There are in fact heavy contradictions that mark contemporary Italy. To start from a given macroscopic see whole areas almost uninhabited, with the abandonment of many countries, especially in the hills of the Apennines. The consequences are large agricultural areas have become unproductive. There is no supervision on the rivers and streams. Consequently grow landslides and other environmental disasters.

In Italy more than 60% of the population lives in the suburbs, very different in economic, social, demographic, and also for urban and architectural levels of degradation. Here we find, especially for young people and new families, a shortage of housing. There are data that show the picture of the situation and should give us pause. In Italy there is a heritage of more than 200 million homes. The inhabitants are about 60 million. This means that each of us would have to take more than three dwellings.

Obviously someone like Silvio Berlusconi is far beyond the three that belong to everyone. It is true that many Italians, especially those who live in large cities, do not even have a house. Many young couples live with their parents while the housing stock and entire small towns are likely to collapse for lack of maintenance.

They appear, however, tentative signs of a turnaround. Not only older people leave the big cities to the high cost of living, especially for rentals; smog, traffic, and they return to their countries of origin. Some young people, for the lack of jobs in large cities move in the small villages with new people and remit to the culture abandoned countryside. Among the most interesting experiences of this turnaround we point out some emblematic cases as two small villages that are Santo Stefano di Sassanio and Solomeo.

Sessanio, a small town not far from L'Aquila in Abruzzo, probably derives from Sextantio, a small Roman settlement located near the present town. The first news we have in 760 A.D. The work of the monastic orders has allowed an increase in arable land, the repopulation of the areas at high altitudes, the emergence and consolidation of fortified villages, safer from marauding barbarians because in an elevated position. At the beginning of our millennium the ancient village is having a rebirth, thanks to the will of the few young people remained or returned, tourism crafted by the will of an entrepreneur: Daniele Elow Kihlgren, of Swedish origin, who bought a part of the village to realize ourselves a popular hotel. This Act has attracted the interest of other investors who are restoring other homes and are developing considerably different activities in the area, starting with agricultural crops quality.

The restoration has respected the architecture of the old houses, the friezes, the angles, the paths, the history and the tradition. The houses are built of white limestone darkened by time as in the Middle Ages. Even small shops, finally reopened, are built into the rock and triumphs over all the amazing view of the Gran Sasso. The new digital technologies make it possible to connect with the world. They are also reported numerous specialties qualifying point of the Mediterranean diet. This experience speaks to us of the beauty of the landscapes spread, the quiet of the ancient villages pedestrian, the charm of the place populated by the generations that preceded us. Da nobis hodie incantum quotidianum reads an ancient prayer and is known as the spell is that useless thing that is essential as bread.

Solomeo is a small village in the municipality of Corciano, near Perugia. Populated by 436 inhabitants in 2001, it is perched on top of a hill at 273 m. above sea level. Date back to the third century BC of Etruscan artifacts found in the area while the name evokes the Etruscan god lumn hence then San Lumeo with whom he began to identify the place.

Here is a sort of leading textile mills in Cashmere: the Brunello Cucinelli SpA that has industrial plants in the valley, while in the village, in addition to the showroom of knitwear, has developed, thanks to the presence of buyers, also the armature linked to gastronomy and tourism in general. A Solomeo Cucinelli has built, restored and made many cultural plants. He realized the Arts Forum, the Peace Square, the Theater, the Library and the School Aurelian, and sewed up the wounds that time had dealt with the old houses and streets, in keeping with the teachings of John Ruskin. The whole area has been turned into a sort of garden spread in tune with the genius loci, or listening to the voice of the spirit of the places that involves constant physical contact with the land and the landscape to create a continuity between man and nature.

This entrepreneur wanted to evoke the spirit of the medieval building site that is the basis of the restoration of the village, headquarters, turned into a place for the soul, for the mind, for the study and culture. Later

also the knitwear factory was installed in a large industrial factory that was abandoned in the valley at the foot of the village. Here large windows contact the inside with the outside, opening up a view of the most beautiful landscapes framed by trees and flowering shrubs and garden visible from all jobs. Operate in visual contact with the surrounding nature, see their houses while you are at work is intended to make less repetitive and more human labor.

And 'in courses carrying an agricultural park of about seventy hectares of land intended for gardens, vineyards, olive groves and orchards spread like trees. Various crops such as wheat, corn, sunflower. Where we cultivate the land with respect for nature and the products are for local consumption in company canteens and for working families.

As to the second aspect of our communication I want to point the intervention at the margins of a common agricultural essentially of Tuscany. It is San Casciano Val di Pesa, which has 17,247 inhabitants in the metropolitan city of Florence, on the hills dividing the river valleys of the Pesa and Greve. Crossed by the Via Cassia, 15 km south of Florence and about 45 km north of Siena, it is in the Chianti area, known for its production of wines, extra virgin olive oil, and agricultural products in general. An agricultural land with the presence of Officine Graphics Stianti, a very important in the economic of the City.

The project, designed by architects Nataliani Associates, as the next, over an area occupied by equipment for the work they have fulfilled their purpose and are precious reserves of space on which the city can expand without taking up new ground.

The historic building has been renovated internally while maintaining the original layout. On the ground they are placed against the cellars and the connection to the car park of the new residences. 152 apartments overlooking the street that develop on parallel dimensions and degrade downstream, as an addition of the historical nature. At the first level we have commercial spaces while upstairs apartments of various sizes. On the ground floor the building is crossed by a public path that passes through the courtyard and on to the lower level of the sector while the underside we find parking. On the ground floor, separated by private courtyards connected by pedestrian public, there is a dense connective tissue of relational spaces.

To the northwest of 's-Hertogenbosch, between the villages of Engelen and Bokhoven, where the Meuse, the river makes a large curve and a canal connects it to Dieze, stretches over an area of about 200 hectares, green and flat, torn waters and protected by a levee built by the new settlement Natalini Associates in Holland.

The general plan for Haverleij, developed by Sjoerd Soeters and a group of landscapers, invented a new land with bars of wood, chess trees, triangles and green circles, between nature and artifice. The plan was a great golf course with 27 holes, an area in the forest, a nature reserve and about 1,000 houses, 600 of them gathered in a village on the other closed and divided into groups of 60, called "castles", positioned the nodes of a regular grid superimposed on the new landscape.

A Natalini was asked to design a "castle" in a pond where shooting a model ideal for groups of houses: a fantastic car metaphor of separation between an external (chaos) and an extension (cosmos). To protect the quality of the city and live. The designer has transformed diagrams of castles and fortifications in mixtilinear a path on which they organized a door and a lunette, a semicircular rampart and walls with corner towers. The houses are set against the fortifications and gradually have conquered and replaced with a fence to get enchanted with the water out and a court-garden inside. The houses are of bricks of different colors: darker parts in contact with the water, the more red in the upper ones. The gates and towers are made of stone, with some walls covered with zinc, as all covers.

As recalled by Le Corbusier "Architecture is a fact of art, a phenomenon of emotion, beyond the issues of construction, beyond. The construction is to hold; the architecture is to move. The excitement is when the architectural work sounds you like to pitch a universe of which we suffer, we recognize and admire the laws. When you reach certain relationships we are captured by the work. The architecture is' relationships' is' pure creation of the spirit. "

It seems that everything is shown from the proposed actions.

康胤
Kang Yin

中国美术学院建筑艺术学院副教授
1988年　毕业于浙江工业大学土木工程系
1988年　宁波市镇海建筑设计院工作
1999年　毕业于东南大学建筑系，获建筑学硕士学位

主要作品
1991年　宁波市镇海南街市场
1992年　宁波市镇海棉纺织厂厂区规划及单体建筑设计
1995年　宁波市镇海龙赛中学科技馆建筑设计
1996年　宁波市北仑地税综合楼建筑设计
1997年　江苏江浦高级中学校园规划及建筑设计
1997年　南京九华山公寓建筑设计
1999年　宁波市北仑残疾人联合会综合楼
2000年　杭州凤栖花园环境景观设计
2001年　安徽青阳宝灵观文化休闲商城
2001年　湖州嘉业阳光城环境景观设计

Professor of China Academy of Art.
1988　graduated from Department of Civil Engineering of Zhejiang University of Technology
1988　worked at Ningbo Zhenhai Architectural Design Institute
1988　graduated from Department of Architecture of Southeast University with Master's degree in Architecture.

Main works:
- 1991 South Street Market of Zhenhai, Ningbo
- 1992 factory area planning and monomer building design for Zhenhai Cotton Mill, Ningbo
- 1995 building design for science and technology museum of Longsai Middle School, Zhenhai, Ningbo
- 1996 building design for comprehensive building of Beilun Land Tax Bureau, Ningbo
- 1997 campus planning and building design for Jiangpu High School, Jiangsu Province
- 1997 building design for Jiuhuashan Mountain Apartment, Nanjing, Jiangsu Province
- 1999 comprehensive building of Beilun Federation of the Disabled, Ningbo
- 2000 environmental scenery design for Fengqi Garden Apartment, Hangzhou
- 2001 Baolingguan culture leisure mall, Qingyang, Anhui Province
- 2001 environmental scenery design for Jiaye Sunshine Town, Huzhou

城市建筑"遗产"再生

康胤

【摘要】以浙江省湖州市南街为例,说明在中国城市发展过程中,如何通过对现有城市街道空间的整体设计,将城市街道建筑外立面进行统一的改造,在提升城市形象的同时,使城市的历史遗产得以保护。

Abstract] South Street in Huzhou City of Zhejiang Province shows how to complete the unified transformation of building façade among the street through the overall space design of existing urban street in the process of urbanization in China, so as to protect the historical heritage of this city while improving the image of city.

"在2010年第3届国际绿色建筑和节能大会上,时任中国住建部副部长仇保兴发言中提到:我国是世界上每年新建建筑量最大的国家,每年20亿平方米新建面积,相当于消耗了全世界40%的水泥和钢材,而只能持续25~30年。而资料显示,英国建筑的平均寿命达到132年,美国的建筑平均寿命达74年。

每万平方米拆除的旧建筑,将产生7000~12000吨建筑垃圾,如此短寿的建筑每年将产生数以亿吨的建筑垃圾,给中国乃至世界带来巨大的环境威胁。"(引自中国新闻网2010年4月5日)

针对我国建筑20~25年的平均使用寿命的说法,意味着城市中建于20世纪80年代、90年代初的建筑都已超过了平均寿命,都已成为随时可能被拆除的城市建筑"遗产"。

自20世纪80年代起,基本解决了温饱问题的中国城市,开始了第一轮大规模的城市建设,直至今日我国的城市仍像一座座巨大的工地,城市在"拆"与"建"的循环中不断演变,从拆除古建筑到大量建成仅二、三十年的"新"建筑被夷为平地,在产生巨量垃圾的同时一个个城市记忆也被瞬间抹去。分析那些"新"建筑被拆的原因不外乎三种:第一类是因建造时施工质量问题,短时间内就成为危房而被拆除;第二类则是因城市规划的因素造成规划用地性质改变而被拆除,或因交通规划原因造成道路拓宽而被拆除.第三类也是被拆除量最大的一类,虽然我国有经济、实用、美观的建设方针,但是二十多年前对囊中羞涩的业主而言经济实用才是重要的,而当年的绝大部分设计师又未受过正规的建筑学训练,缺乏经验的设计师在大量的设计任务面前通常是生搬硬套的拼贴。随心所欲,甚至变形金刚式的建筑统治了城市的各条街道(图1、图2)。进入到21世纪,随着经济的发展,这种杂乱的建筑形象已难以满足人们对城市形象的美观要求,在商业利益、行政利益的共同驱动下,"新"建筑被大量拆除。过度的拆、建在对环境造成不可估量的破坏同时,对城市文脉传承也造成无法弥补的损失。

图1(左),图2(右)

近十年来，随着全社会环保意识的不断加强以及市民维权意识的增加，城市中拆迁难度越来越大，整街拆除在很多城市已难推进，建筑立面改造应时而生。以浙江为例，从2006年杭州市区的庆春路、凤起路、体育场路，到2008年湖州市区的南街、北街、人民路，再到2010年安吉县城的天荒坪路、浦源大道等，城市沿街建筑立面改造与交通、景观整治尝试从省城向地级市、县城推进，数以万计的"新"建筑得到了改造，城市形象也有了显著提升，城市建筑"遗产"得以保留。

2007年底受湖州市政府的委托，本人承接了湖州江南工贸大街建筑立面改造工程设计。经过初步了解，原来这条大街的街名在1999年政府通过拍卖把冠名权卖给了企业，它的原名叫南街，在古城这可是一个沿用了千年的名字，南街一直是水乡古城繁华的象征。进入调研阶段后发现仅做建筑立面改造，这些立面被改造的城市建筑"遗产"的形象是无法有特别大的提升的，也难以让城市风貌有质的飞跃，还是无法摆脱随时被拆除的命运。经过与项目委托方多次沟通得到认同之后，确定项目由"江南工贸大街建筑立面改造设计"变更为"湖州南街城市街道空间整治设计"，提出了站在城市街道规划的视野下，以全面调研分析为前提，制定出详尽的街道规划方案，然后对道路、景观与建筑立面整治设计同步推进，全面提升街道空间形象，从而使城市风貌发生质的飞跃，城市建筑"遗产"得到有效保护（图3）。

图3

针对建筑立面整治，综合各类调研信息，把沿街建筑分为清理、小尺度整治、大尺度改造三大类，对

图4（左），图5（右）

图6（上），图7（中），图8（下）

建筑立面品质较好，建成时间不久的建筑只进行墙面清理、空调位整治；对建筑造型完整，局部杂乱的建筑仅作局部小尺度整治。最典型的案例是湖州市一医院的医技楼，建于1986年，框架结构，马赛克墙面，建筑造型完整，比例协调，建筑造型、材料有明显的年代特征，是20世纪80年代难得的优秀设计作品，在我们的强烈要求下仅作了底商局部改造，为城市、为街道留下一个令人回味的记忆（图4、图5）。面对千家万户的大尺度改造，调研的作用尤为重要，整治方案必须从城市、街道、建筑自身、业主、政府包括设计师在内的各方诉求中寻找最大公约数，达到多方共赢的目的。南街立面整治涉及28幢建筑、上百家商户、上千家居民，其中有12幢建筑需要大尺度改造，政府部门虽已做了大量细致有效的工作，也得到了绝大多数商户、住户的大力支持，但施工中还是遇到了个别业主的坚决抵制，设计工作只能随时进行调整，也使整个工期从半年延迟至一年才得以完成。最典型的案例是湖州经协大厦，建于1990年，框架结构，安全性良好，底商办公楼，变形金刚式的建筑造型，身处黄金地段却是城市中最低端的商务楼，已有开发商看中了这块寸土寸金的宝地，建筑已成为名符其实的城市建筑"遗产"，调研之初4家业主各谋其政，各方要求根本难以统一，项目进度受到了很大阻力，经过六轮方案修改，八轮正式协调会后得以通过。半年后焕然一新的商务楼得到了各方的一致好评，办公楼租金直接上涨了两倍（图6、图7、图8）。经协大厦南侧公寓楼改造也让建筑从破旧的集体宿舍成为了高端公寓（图9、图10、图11）。

施工过程中，道路整治、立面整治同时进行，景观整治则是待前两项工程进入收尾阶段后全面铺开。统一在城市街道空间整治原则下的南街，建筑形象有了重大改变，街道面貌焕然一新，在同期开始整治的4条街道中一枝独秀，也为后续街道整治树

图9（左上），图10（右），图11（左下）

立了标杆。随后我们又对城市的东街、人民路、凤凰路等9条城市中心干道进行了整治设计三年改造使湖州城市风貌有了真正质的提升。2010年后又对安吉县城、南浔镇区6条道路进行了整治设计，涉及建筑立面改造452幢，大量待拆的城市建筑"遗产"获得新生（图12、图13、图14、图15），还有待拆的城市节点得以保留。在湖州凤凰路整治过程中，从城市规划及历史信息的调研中发现，凤凰路与青铜路十字路口的环岛绿地，建于1996年，圆环直径60米，有一组湖州开发区主题雕塑，景观效果好，地标性城市节点，但因环岛汽车通行率的限制，政府已经决定拆除。面对这么重要的城市节点，经过有力的数据分析，并提出了极具说服力的整治方案，终于说服政府，保留了湖州仅存、江南少见的市中心环岛绿地，为城市留下了一个值得记忆的空间节点。道路整治中针对有些建筑退让道路较大而成为停车场，造成机动车与行人混行，人行道严重破损的现状，在湖州凤凰路、安吉天荒坪路、南浔人瑞路改造中，经充分调研，权衡各方诉求，创造性地提出了：拓宽非机动车道，让机动车停放在非机动车道上，在兼顾商家利益的基础上，有效解决了停车、行人、非机动车三者之间的矛盾，取得良好的效果（图16、图17）。

　　成功的城市街道空间整治是要以城市总体规划为基础，制定详细规划，建立街道空间整体整治模型。

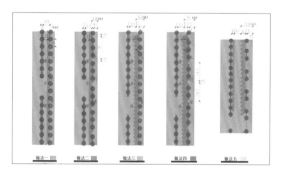

图12（左上），图13（右上），图14（左中）
图15（右中），图16（左下），图17（右下）

详尽的调研分析是改造成功最为关键的环节，后续的具体设计虽有许多细节需要处理，许多技术难题需要解决，但也就是顺势而为的工作罢了。整治后的建筑形象、街道形象、城市形象的全面提升，自然是水到渠成了。如此众多的城市建筑"遗产"得以保护再生才是最值得设计师欣慰的所在。

阿哈姆德·拉希德
Ahmed Rashed

埃及英国大学建筑工程系教授
可持续性和未来研究中心创始主任
埃及工程师联合会顾问工程师
建筑学博士，约克大学，英国

2013年　至今任AGTBE（驻埃及英国毕业生和实习生协会）主席
2011年　任埃及建筑师学会理事会成员
2004—2008年　任埃及曼苏拉大学建筑系负责人
1984年　在埃及艾斯尤特大学获得学士学位
1990年　在埃及艾斯尤特大学和沙特国王大学获得理学硕士
1995年1月　在英国约克大学获得建筑规划与保护的博士学位，致力于在保护埃及卢克索地区文化遗产方面提高公众参与度。

他还曾在埃及、沙特阿拉伯和阿联酋的几所不同的大学任教。Ahmed Rashed教授的研究领域主要包括建筑历史及理论（尤其专注于古埃及法老历史），建筑规划及可持续发展和绿色建筑。此外，Rashed 教授也是推动埃及文化遗产版权保护活动的主要领导者。

Professor Architectural Engineering, The British University in Egypt.

Professor atArchitectural Engineering Dept., the British University in Egypt, Founding Director, Centre of Sustainability and the Future Studies
Consultant Engineer:According to Engineering Syndicate (No: 2531/2)
Ph.D. In Architecture University of York, UK 1/1995
Prof. Ahmed Rashed is a professor of history and theory of Architecture (especially for Pharaonic Egyptian History), Planning, Sustainability and Green Architecture.
Also Prof. Ahmed Rashed is the leader of the Egyptian Heritage Copyright petition

The president of AGTBE | The Association of Graduates & Trainees from Britain in Egypt from March 2013
Member Board of The Egyptian Society of Architects June 2011
The Egyptian Syndicate of Engineers
The Egyptian Society of Engineers
The Arab Archaeological Union
The AmericanResearchCenter in Egypt
The Egyptian Exploration Society
The Tohoti Committee (NGO)
The Environmental and Health law Committee (NGO)

文明将孕育出新的文明
——卢克索地区文化遗产及可持续发展面临的挑战

阿哈姆德·拉希德

一、摘要

卢克索作为埃及古都及人类文明的中心有着数千年的历史，世人渴望了解它也想通过它寻找人类历史发展的轨迹。可以说，卢克索就是埃及的缩小版，同时它也面临着现在埃及所遭受的全部挑战。卢克索因其丰富的文化遗产，多彩的文明及潜在的财富而成为了一个独特的地区。因受当今埃及政局的影响，曾作为旅游之都的卢克索，旅游市场客源萎缩，经济下滑，人民也因不能满足日常所需而造成生活质量严重下降。除此之外，卡尔纳克神庙，卢克索神庙，帝王谷及其他著名遗产的建筑物都遭受到了不同程度的损毁，所以，为了能使埃及的文化遗产得到更好的传承及实行可持续性的发展，卢克索人及其他埃及人民，乃至国际社会都有义务为此做出积极的改变。这篇文章的目的是通过研究这些不同的遗产，综合卢克索和埃及青年及国际社会的意见，为每一处遗产找到特征鲜明的商标。基于这一目标，本文也将对在商用或盈利环境中使用埃及文化遗产的复制品是否侵犯埃及文物版权而作出讨论。

二、埃及文化遗产的版权保护

之所以要将这个十年之前的议题再次提起，是因为现在我们不得不面对的一个事实是埃及的文化遗产几乎在世界各大博物馆都存放着复制品，这从某种程度上意味着这些复制品侵犯了埃及文化遗产的知识产权。

当然，人们对如何处理这项议题也存在着长期的争议，有些人认为这些在世界各大博物馆或是展馆展出的复制品能够向世人介绍埃及，是一个非常好的宣传方式。今天的埃及正寻求一切机会重塑埃及精神，也希望从这片土地中寻找到新的经济发展途径，从而更好地建设新埃及。

假如埃及的旅游业已陷入困境而他们也正在寻求不同的经济发展途径，那么我们就有必要找到一种新的经济来源，促进埃及发展。

本文将重点对埃及文化遗产在版权保护方面一直存在争议的问题作出讨论。其中，本文还将以埃及卢克索与拉斯维加斯卢克索之间的对比为例，阐述许多地区在使用埃及文化遗产的复制品时并未对其所侵害的版权做出任何赔偿（在法律及文物版权共同协议限制范围之内）。虽然在过去的十年中我们为保护埃及文化遗产的版权采取了许多措施，但是我们仍需要好的建议及各方的支持。现在，有两个关于文化遗产版权保护的提案正慢慢浮出水面：第一，是开展一项以"怎样在国际合作中获得已用作商用的埃及文化遗产（及其复制品）的知识产权"为主题的比赛；第二是为埃及的文化遗产创立品牌。

三、背景

卢克索（即古希腊的底比斯）作为世界著名文化遗产的宝库及埃及旅游的重要景点之一，它有着5000年的历史。在这种长期的历史发展进程中，其周围的自然、社会及政治环境都在随之改变，所以这也使卢克索地区文化遗产保护成为了一个相当艰巨的任务。由于在过去的几十年中人口的高速发展及每年有成千上万的游客来卢克索，这些因素也让卢克索地区的文化遗产遭受着比几千年的侵蚀还多的破坏。埃及方面曾为保护卢克索地区的文化遗产作出呼吁，也开展了研究及规划工作，但遗憾的是，现实中的保护工作却一直让人沮丧。

另一方面，一个模仿卢克索地区文化古迹的建筑物出现在了美国拉斯维加斯卢克索酒店的建造项目中，这也是典型的美国梦的象征。卢克索酒店是完全依照埃及金字塔的外形进行建造的，大约有30层，其中能容纳4400个房间及约11000万平方米的赌场。在该酒店的各个楼层中也充斥着各种仿制的"埃及"景点及建筑结构，例如法老墓及法老与王后像的复制品。使用那些打着"文化遗产"名头的复制品来促进旅游业的发展，确实是非常有效的商业手段。但讽刺的是，这样的做法使得"卢克索"这个名字在今天已成为了美国拉斯维加斯卢克索的代名词，而非真正的埃及卢克索！

　　本文的目的有两个方面：第一，本文将在旅游管理及文化遗产保护方面对古迹的原版及其复制品进行比较研究；第二，对文化遗产的复制权作出讨论，并研究其影响。具体来说，本文将讨论以下几个论点：

　　对埃及卢克索和拉斯维加斯卢克索来说，他们各自对旅游业有着不同的定义。

　　真正的文化旅游包括休闲旅游，而休闲旅游也可以利用文化旅游促进其发展。

　　旅游管理是一门艺术，它的主要思考对象是如何将人引入到一个与自己生活区域完全不同的地方。

　　在全球化的时代，对于埃及卢克索及其复制品——拉斯维加斯卢克索之间的讨论应该将重点放在复制权和文化产权（知识产权的同义词）上。

四、旅游业及文化遗产市场

　　旅游业是一种主要的经济发展方式，也是世界上最大的产业之一。尽管旅游业并不是在历史文化名城中唯一的活动形式，但是它对文化古迹保护的影响还是独一无二的。旅游业也是一项比较大的出口产业，因为每年都有数以百万计的人在旅游市场上消费，所以它也是一种赚取外汇的途径，可以说，在全球外贸领域中，旅游业占据着最大的一部分。同时，旅游业常被认为有能力促使经济在短时间内实现增长[1]，这也表明了它能为国家经济作出重要的贡献。

　　但是，定义旅游业是非常困难的，因为他并不能简单地依照标准产业分类法（SIC）进行分类。旅游业的特点，不在于其产品，而在于产品的购买者，也就是"游客"。大多数对旅游业的定义是指专门为游客提供游览、住宿、餐饮、购物、文娱等环节的综合性服务行业[2]。除此之外，旅游还分为国际旅游和国内旅游。联合国就将国际旅游定义为"一个国家的居民跨越国境到另外一个国家的旅游活动并至少要在另一国待满24小时以上"。而国内旅游指的是国家内居民离开长住地到国内另一地方去进行的旅游。根据不同的旅游访问目的，人们对旅游有了更为精确的定义，即商务旅游，休闲度假旅游及文化旅游等。因此，"发展旅游能附带提升经济的发展"这句话是非常正确的。[3]

　　对于埃及卢克索及拉斯维加斯卢克索来说，"文化/历史遗产"是吸引游客的重要手段。它是一笔社会财富和一种政治及经济资源。因为"文化/历史遗产"其本身就具有价值，所以文物古迹可以被收藏、保存及作为展示。一个国家的文化遗产不仅代表着这个国家的历史，还"充分展示了其民族的独特性"，因此，它能增强民族团结，提高民族自豪感，更是展现国家形象的一张重要名片[4]。此外，"文化/历史遗产"作为一种经济资源，有着多种使用形式。那些所谓的"文化/历史遗产"旅游其实就是"以文化遗产为名实而进行买卖商品和服务的商业活动"。虽然旅游及休闲服务在旅游行业中占据着非常重要的位置，但是制作及销售文化遗产产品也同等重要。文化遗产可以被用来打造地区形象，以达到宣传目的。所以文化遗产旅游指的是"要在文化遗产的基础上发展旅游业，要提供以文化遗产为核心的旅游产品，要把文化遗产作为刺激消费者进行旅游消费的主要因素"[5]。"文化/历史遗产"旅游能让人们通过游览、参观文物古迹及参加各式活动对旅游地的历史及现状有更为真切的感受。它还包含文化、历史及自然资源。

　　历史遗产或文化旅游早就成为了旅游的一种形式，例如早在19世纪时期外国人在埃及进行的探险考察活动。但从另一方面来说，文化旅游在今天仍占据着埃及旅游市场最大的份额，卢克索就是主要景点。在世界的另一端，对于美国50个州中的47个州来说旅游业只是他们追求经济利益的手段[6]。尽管美国与埃及在开

图1：左图为埃及卢克索神庙，右图为拉斯维加斯卢克索酒店的入口

图2：左图：克隆的历史遗产，美国拉斯维加斯卢克索酒店

右图："借来的"历史遗产Dandor神庙，纽约大都会艺术博物馆

展旅游业的形式及目的上有较大差别，但是美国人却以"借来的文化遗产"（例如Dandor 神庙）和"克隆的文化遗产"（拉斯维加斯卢克索酒店）为名，促进当地旅游业的发展，甚至想以此打上文化旅游的标签。（图2）

当然，我们也必须承认其他类型的旅游方式也非常重要——假如其他类型的旅游方式并不像文化旅游与历史遗产及保护之间有着非常紧密的联系，换句话说，"在旅游城市中，历史文化遗产才是唯一的旅游资源。"[7]

五、埃及卢克索古迹管理及文化旅游

对于那些向往了解埃及古迹的文化旅游者来说，卢克索是必经的一站。卢克索地区的气候条件（尤其在冬季）通常以温和、干燥及晴朗的天气为主。各类游客不仅可以在卢克索感受到底比斯山及尼罗河的自然风光，还可以在河谷地区感受到当地的乡村生活。

在古埃及新王国时期，一座当时世界上无与伦比的都城底比斯（现卢克索）雄踞在尼罗河河畔。底比斯的东岸（日出）是当时古埃及的宗教、政治及进行日常活动的中心，而西岸（日落）则是法老们死后的安息之地。所以，卢克索古迹的历史可以追溯到公元前3000年的法老王朝时代。卢克索拥有的世界文化遗产包括帝王谷、帝后谷及贵族墓葬群。另还包括孟农巨像、卡纳克神庙（埃及最为雄伟的法老神庙）及卢克索神庙这些建筑杰作。这些文化遗产的存在展现了人类的早期文明，也证明了人类取得的文化成就。这也是世界各地的游客对埃及文化遗产着迷的原因。

在卢克索人看来，旅游业一直是当地主要的经济活动，因为它能创造更多的就业职位及商业机会。但是，当因地区及政治争端而引起社会动乱时，当地的经济也会随之受到严重影响。

这些著名的文化遗产吸引着全世界越来越多的游客来到卢克索。但这种持续增长的文化旅游对卢克索地区的文化遗产来说却是一种威胁。例如大量的游客会对陵墓中的壁画产生不利影响，使其变得不易保存。又比如持续攀升的游客人数及人群拥挤会直接降低游客对神庙的文化体验。同时，一些以商业、考察、访问为目的的短期旅行并不能为当地经济带来很大的好处，更大大减少了游客在选择参观景点时的灵活性。

因此，卢克索的旅游业现阶段面临最大的挑战是要在保证一定数量游客的基础上，确保埃及的文化遗产不会因旅游业而遭到破坏。除此之外，旅游业是一个复杂、快速变化及富有竞争力的行业，所以卢克索应努

图3: 卢克索, 左图: 埃及
右图: 拉斯维加斯

图4: 卢克索 (狮身人面像大道), 左图: 埃及
右图: 拉斯维加斯

力作好旅游市场的研究及规划以跟上世界潮流。虽然旅游业普遍被认为是一个无污染的行业（无烟囱或危险化学品），但是它对社区的基础设施有较高的要求，例如公路、机场、供水、警局及消防等公共服务，而建设这些基础设施却会对文化遗产产生不利影响。

六、拉斯维加斯卢克索：旅游管理及文化遗产消费

拉斯维加斯是世界发展最快的城市，也是让人充满幻想的城市。旅游业的持续增长催生了拉斯维加斯的酒店业市场持续扩大。其中的一个最大的项目便是拉斯维加斯卢克索酒店。

卢克索酒店建于1993年，其可观的建造额（3亿美元）也使这个项目注定不平凡。在项目竣工时，这个具有30层的埃及主题酒店，容纳了4400间客房，一条室内尼罗河及10条接驳船，并且还再现了法老图坦卡蒙的陵墓。除此之外，酒点的赌场还设置了数量惊人的赌桌及老虎机。

在一篇短文中描述整个卢克索酒店就是一个城市项目，这个由混凝土及玻璃建造的金字塔建筑非常吸引人的眼球，人们可以选择在此住宿、娱乐及赌博。酒店中设置了连续的环形走廊式阳台可俯瞰中庭，每一处皆比底层高2米。卢克索酒店的中庭也可以说是全世界最大的。考虑到整个酒店的建筑结构为金字塔型，工程师便将酒店中的电梯以39度角向上移动。酒店的正门是一个钢结构的埃及狮身人面像复制品。尽管整个狮身人面像的比例并不正确，但它仍非常壮观。因为真的狮身人面像要与金字塔的比例相协调，所以卢克索酒店的狮身人面像略大于真的狮身人面像。相比之下，拉斯维加斯卢克索酒店的金字塔却小于埃及吉萨的大金字塔，但酒店金字塔的高度仍受到了联邦航空规则的限制，因为他们仅离机场一千米。狮身人面像下的景观也使用了埃及当地的植物，让人有一种真正来到埃及金字塔的感觉。当沿着卢克索酒店的人行道行走时，你将会首先看到一条周围全是狮身人面像（复制品）的大道，这也与卡纳克神庙及阿蒙神庙的景象相类似。（图4）

该项目的目的是让游客在进入酒店之前有一种赏心悦目、身临其境的体验。酒店中使用的雕像都精确复制了埃及的原始作品，在景观中大量种植棕榈树不仅让整个区域看去更像沙漠，也为游客在游览时提供了遮蔽处。酒店还建造了一座与埃及阿蒙神庙前类似的方尖碑。（图5）

每个房间都配备了具有埃及风格并以象形文字装饰的衣橱，还包括一个衣柜、抽屉和一个松下电视。墙上挂有克莉奥帕特拉（埃及女王）壁画的复制品。酒店的浴室要比房间还大。许多人都很好奇房间的形状是否是小的金字塔型，实际上他们呈矩形状，并且窗户所在一侧的墙壁向内倾斜。上面的房间都是通过名为倾倒器的工具连接的。普通的垂直升降机除设置在中庭的中心区域外都会以最短的距离进行上下移动，因此在

卢克索酒店的四大角落都设有一个倾斜的升降机并将电梯箱以39度的角度上下输送。每一个角落的电梯都会在特定楼层停靠，所以住宿者必须知道哪一部电梯会将他们输送到想要到达的楼层。

卢克索酒店的赌场并不大，可能是因为整个赌场是以环绕中央的方尖碑而布局的吧。一些具有主题性质的老虎机作为补充被摆放在各处，例如"法老的宝藏""金色眼镜蛇"和"帝王谷"。正如所看到的，卢克索酒店的设计者们只是设计了一个以

图5: 卢克索（入口的方尖碑），左图: 埃及
右图: 拉斯维加斯

埃及为主题的主题酒店，仅此而已。相对于其他的赌场，卢克索酒店的赌场是比较安静的，并且投币口的响铃声也非常轻，可能是想让金字塔酒店更像法老的陵墓吧。商店也销售着各种"埃及商品"（"这里有你所需的关于尼罗河的任何商品"，包括各种甜食，木乃伊橡皮糖）。酒店的餐厅也包括伊西斯（古埃及女神），圣海房及金字塔咖啡厅。

酒店的规划者又进一步对埃及主题进行了探索并开发了数个景点，例如"与考古学家一起游尼罗河"，有关埃及文物及历史的3D电影及埃及卢克索最为著名的"法老图坦卡蒙之墓及其博物馆"（现已不在酒店展出）。这些"法老图坦卡蒙的陵墓及宝藏"的复制品给人的感觉就真的像是在"1922年的11月被英国考古学家霍华德·卡特发现"。在游览过程中，酒店还会播放磁带引导游客通过展览室。

总的来说，拉斯维加斯卢克索酒店用了大量的资金对埃及卢克索的文化遗产进行复制以促进当地博彩业及旅游业的发展。卢克索酒店为那些没办法来到埃及的人们提供了一个机会（每季度10000人）。换句话说，卢克索酒店在以发展博彩业的前提下，开发了独特的"文化（文化遗产）旅游产业"。

但从另一方面来说，开设赌场在拉斯维加斯被认为是合法经营，并且它会一直存在。赌场的全年收入要高于观看电影、体育赛事及主题公园游览这三者加起来的收入。即便如此，也只有约30%的美国成年人去过拉斯维加斯。

七、文化遗产的真实性及版权

文化遗产是一个非常广泛的概念，包括自然遗产传承和文化遗产传承[8]。文化遗产作为建成环境的一部分，为使用者提供了特殊价值，所以在特定的城市环境中它是一种主要资源。而在文化背景下，文化遗产可分为物质文化遗产及非物质文化遗产，例如手工艺品、纪念碑、历史遗址、建筑、哲学思想、传统习俗、庆典仪式、历史事件、独特的生活方式、文学作品、民俗活动或教育思想。

对于大多数商品来说，专利所有权可以保证该商品在市场中长久生存并繁荣发展[9]。这些专利所有权的性质也会根据不同种类的文化遗产进行变化。假如文化遗产中附着着特定价值，那"文化遗产的占有权"就非常有必要存在。也许很多人都想问所谓的特定价值到底指的是什么，是谁将这些特定价值附着在文化遗产之上的，文化遗产的占有权到底指的是哪一种。通常，"文化"这个词多被用来描述这种特定价值，并且我们俗称的"文化遗产"仅仅只是真正文化遗产的一部分。

管理文化遗产的首要目标是将保护文化遗产的重要性及必要性的知识传递给当地居民及游客。"文化"中的各种具体表现形式，比如考古遗址、人文历史、传统习俗及生活方式，之前常被视作推销文化旅游的主

图6: 卢克索 (法老图坦卡蒙之墓), 左图: 真品
右图: 复制品

图7: 金字塔, 左图: 埃及吉萨
右图: 美国拉斯维加斯卢克索酒店

要工具。文化/历史遗产可谓是人类生活中不可替代的资源。

埃及作为"世界文化遗产"的一部分,其文化遗产的真实性是毋庸置疑的。因此,在《国际文化旅游宪章》[10]的原则4.2中提出"应该尊重一些社区或土著居民要限制或管理通向一些文化活动、知识、信仰、活动、人造物或场所的通道的需要和愿望",宪章还在原则4.1中提出"旅游区的地方权利和权益、古迹财产拥有者和相关的、对土地和重要遗址拥有权力和义务的土著居民,应该得到尊重。在旅游背景下,他们应该参与到为遗产资源、文化活动和当代文化表达制定目标、策略、政策和条约的工作中"。

从另一方面来说,如果将这些文物作为艺术作品,那么它们就很可能拥有版权。因为"只有版权的著作人才有法律权利对其作品进行复制,表演或是展出,也可传播作品副本或将作品进行改动;任何人或组织对作品采取未经授权的行为即是'版权侵犯'。这种行为是可以向联邦法庭提起诉讼的"。[11]

这项主张可以通过一些法律条例进行着重展开,例如希腊法律明确指出"未事先向希腊文化部申明就使用希腊文化遗产元素作为商标,图案或样品的行为是严厉禁止的"[12]。此外,类似于拉斯维加斯卢克索酒店的这类项目不能因"合理使用"就免除其的侵权行为[13],原因有二:第一,这是在纯商业活动中使用埃及文化遗产复制品。比如获取经济财富是开发旅游及休闲旅游的主要目标,这样的行为也潜在的限制了埃及旅游业的发展。第二,对埃及文化遗产的复制数量巨大。拉斯维加斯卢克索酒店使用的各种元素皆来自埃及。不仅只在建筑风格方面,只要是涉及旅游及博彩业的每一个项目都会有埃及元素的出现(房间内部、机器和服务,即使是赌博时使用的代币也充满着埃及元素)。最后,对卢克索文化遗产的重现严重威胁了埃及卢克索的旅游市场,当人们在谈论到卢克索时,人们首先会想到的是"拉斯维加斯卢克索"而不是"埃及卢克索"。

八、图坦卡蒙陵墓的复制墓及竞争案例

介于最近发生的一些事件使得埃及文物复制品及版权保护的事宜又再一次被提起。一个由西班牙、英国和瑞士公司组成的团体在2012年11月建造了法老图坦卡蒙陵墓的复制墓(成本超过四万埃及镑)并打算将此作为礼物送给埃及,但问题是要将它存放在埃及的哪个城市(是选择在卢克索,Harghada或是沙姆沙伊赫),最后经决定将其存放在卢克索。但埃及接受这个礼物也意味着这个公司有权利可以制作其他的复制品模型,并且埃及方面不能制止此公司进行任何的文物复制品交易。

2014年4月30日,图坦卡蒙复制墓正式对外开放,埃及旅游部及文物部的两位部长,卢克索省省长,欧盟驻埃及大使及其他25位外国大使都参加了此次庆祝仪式。从正常的距离观看,复制墓与真实陵墓几乎一模一样。复制墓被放置在卢克索的一家小型的博物馆中,这家博物馆揭示了复原时用到的各项技术手段,

也阐明了保护文物的难点。从那天起,围绕图坦卡蒙陵墓的版权争议一直未曾减少。人们思考的是:是否复制墓真的将会取代真实陵墓接待任何以旅游为目的的访问,是否真实陵墓会因复制墓而丧失其的唯一性,文化遗产的版权与商业使用在何种程度上会有争议,是否埃及文化遗产已全是复制品,现在的埃及为什么不能保留其原有的文化遗产。正是这些声音的出现促使了我们对埃及文化遗产复制品现状的调查并对侵权行为进行投诉,其中几个比较

图8:卢克索之夜,左图: 埃及
右图: 美国拉斯维加斯

著名的例子有"美国拉斯维加斯卢克索酒店""迪拜的法老之乡"及其在中国、日本、美国的几个项目。当我们将拉斯维加斯卢克索(每年有将近五万游客)与埃及卢克索作简单对比时,我们发现了知识产权对于保护我们的文化遗产的重要性。

既然要争取我们的权益,首先要找到一个国家因建筑遗产知识产权的争议而获得赔偿的案例,由此我们可以从这个案例中寻找到解决埃及文物侵权问题的关键,或者可以像建筑师一样提出问题,我就认为埃及的历史遗迹一开始只是想法,后来变成了一项建筑工程,又随着时间推移转变为了古代遗迹。

九、文明将孕育出新的文明

2014年5月21日,在卢克索政府的监督下和埃及大不列颠大学可持续发展及未来研究中心的支持下,一场国际竞赛将会在埃及卢克索举行。这场赛事的主体将会是年轻人,目标是通过这场赛事从各方面提升卢克索地区人民的生活质量,复兴卢克索文明,为下一代的发展打下坚实基础。

这是埃及和卢克索应得的,而且埃及也需要一个标杆,将对未来埃及的研究规划转变成真正的国家政策实施到具体项目中去。但是为什么选择卢克索? 对于这个承载着埃及七千多年的历史及拥有巨大的文化财富和可能性的城市来说,卢克索对第一人类文明之都的称号当之无愧,它也承担着国际旅游城市的责任。

在宣布举办此次比赛前的几个小时,我正在准备发布会所需要的材料,这也让我想起在过去的30年中我一直从事着卢克索文化遗产方面的研究,这些研究饱含着我的抱负、汗水、努力及寻找一切的机会去尝试,当然也会有沮丧与绝望。因此,在比赛发布会之前,我一直都在问自己,为什么选择要在今天宣布这项比赛,这项比赛是否会取得圆满成功还是我们会错失这次机会。

当然,我们也还要考虑举办这次比赛的初衷是什么,它的目标是什么,比赛的管理机制又是怎样的,卢克索对埃及人或是来自不同国家的人来说又意味着什么。

用挑战这个词来描述此次发布会在合适不过了,因为我们要在半个小时内向公众阐明举办此次比赛的原因、方式及时间。

此次发布会也吸引了卢克索当地居民、商界代表、银行、企业、媒体、青年团体及卢克索政府官员(政府官员以前很少有兴趣关注科学研究及青年人的想法)的参加,并且卢克索的政府官员也在此次发布会中着重强调了举办这场国际赛事的目标即是使卢克索地区可持续地发展。

此次比赛的主题是: 为卢克索创立品牌和知识产权保护。卢克索悠久的历史,多彩的文明及其拥有数量庞大的文化遗址都值得让卢克索注册品牌商标,当地的社区也应该着重宣传卢克索的品牌价值。基于卢克

索有着多达15处的文化遗产，例如卡尔纳克神庙及其周围的文化遗产群，卢克索神庙，阿尔卡巴沙路，帝王谷等，我们要对这些文化遗产做好注册商标日期登记，并为它们制作特征鲜明的可应用于市场营销的品牌图标，以保护卢克索文化遗产的知识产权。比赛的第二部分将会讨论全球的知识产权及世界各地利用埃及文化遗产复制品进行盈利的案例。

十、总结

我们可以将城市及建筑遗产的知识产权问题归纳为以下几点：

知识产权为保护产品的创新性已被应用于不同的领域，如果有谁以盈利为目的对我们的自然文化历史及建筑遗产进行复制及进行商业宣传，我们应依照知识产权法向他们要求索赔。我们可以将索赔的资金直接用于遗产的保护及修复，而不需要再等待专项资金的发放及社会团体的捐款。

这不仅仅只限于文化遗产复制品的知识产权，还包括那些已在全球各大博物馆展出的埃及珍贵文物的知识产权（因为我们无法将其索回），例如埃及王后纳芙蒂蒂的头像。对于许多游客来说，能见到纳芙蒂蒂的头像及其他珍贵的埃及文物是非常幸运的，所以在不能向博物馆索回文物的前提下，埃及方面唯一做法便是要求对这些文物保有知识产权。

对于那些拥有文化遗产的国家来说，例如希腊、墨西哥，对其持有的文化遗产保有所有权是为了能更好地保护历史文化遗产的知识产权。

毫无疑问，关于政治及经济方面的真实与谎言都会以概念的形式存在在国际社会的记忆中，记忆是会随着时间而消逝，而人类及建筑文化遗产则是历史事件的见证者，或许在百年之后，他们依然存在。

我们对这些复制品可以在很多方面提出问题，例如在道德层面，利用埃及文化遗产大肆吸引游客，实则发展博彩业，这样的行为是否妥当？又例如在市场营销层面，那些受到法老的遗产启发的各类产品、游戏及使用的货币经过制造再进口到埃及市场，在这过程中埃及是否获得过一点好处？或者中国工厂才是真正的赢家？（我们要求埃及文化遗产拥有知识产权一方面的原因是：当你在网络上搜索卢克索时，你会发现大多数的网站都在讨论拉斯维加斯的卢克索酒店而不是埃及卢克索。）

埃及可以通过共享大陆获得数以万计的利益，假如制定知识产权法律条例还需要一段时间，但是我们可以依靠埃及人民的呼声来唤醒全世界人民的良知。埃及的文明已随着剽窃，忽视慢慢消失在人们的视野中。所以国际社会应对那些利用埃及文化遗产复制品获取利益的公司和机构施加压力，让出他们利益的一部分以为更好地保护埃及文化遗产。

对于拉斯维加斯和埃及卢克索来说，旅游业是两地主要的经济发展方式，很显然，拉斯维加斯在旅游人数及经济收入这两方面比卢克索要好得多，这篇文章的目的也并不是在苛责拉斯维加斯的卢克索项目。但是，这也直接反映了两地在政治及经济方面的稳定性，并且我们也承认卢克索项目只是一个单纯的商业行为，而埃及卢克索的旅游业却牵涉到当地的社会、经济、政治和文化等相关因素。最后，这篇文章旨在突出以下内容：

埃及卢克索，作为文化遗产版权的持有人，应该从拉斯维加斯卢克索酒店及其他文物复制品中获得经济收益，除非这些文化遗产被国际社会公认为其产权是共有的，那么如此，国际社会也应该主动承担起管理这些文化遗产的经济责任，而不仅仅是根据《世界文化遗产公约》只提供道德及技术支持。

The Civilization will be Born from the Womb of Civilization:
Luxor Heritage and the Challenges of Sustainable Development

Ahmed Rashed

1. Abstract

Luxor the unified capital of Egypt and human civilization for thousands of years, studied everywhere, the world yearn her and want to breathe the lines of human history. Luxor is the minor model of Egypt as a country with its whole challenges. It is very unique place with its heritage, civilization and underlying wealth. Luxor people are suffering for not providing the needs and requirements of quality of life. Luxor was affected by the current condition of Egypt now and because it is the capital of tourism all over the world, it was negatively affected and suffered economically and Luxor people suffered too. The unique heritage Karnak temple, Luxor temple, Valley of Kings and Queens, Tut Ankh Amoun and others significant sites physically suffered and the challenges of sustaining heritage and sustainable development must have a mechanism of bottom up approach through the community of Luxor and the Egyptians and the global community to make a positive change The presentation aims to study those different heritage and involve Luxor's and Egyptian youth, the global community in ideas to find a distinctive trademark for every heritage site. Integrating with that objective the presentation will discuss scenarios to claim the Copyrights of Egyptian heritage of using the Egyptian heritage replica in profitable/commercial projects.

2. Egyptian Heritage Copyright

Why we have to re-open the issue has been raised for a period nearly ten years ago and has been decided to freeze not cancel, and today you find a need to discussion. What is the issue: its rights called by the intellectual property rights of the Egyptian architectural heritage in front of each project based on its construction on the Egyptian heritage or original heritage which is replicated in every international museum and we cannot get it back.

There's a Long controversy between the logical and submissiveness and underestimated issue, and claiming that these profitability cloned projects, and the antiquities which in the international museums or the global or exhibitions considered it as a good publicity or promotion for Egypt. Today Egypt are looking for every opportunity to back the spirit and the construction will be by searching for how to finance the construction of New Egypt from the land of Egypt, and Egypt's civilization.

If tourism as a source of financing is in crisis and while Egypt is searching for funds from different sources, there is a need to find new unconventional sources of funding by creativity and logic.

This paper will highlight the different scenarios of the debatable issue of heritage copyright. The dialogue of two Luxors': Luxor Egypt versus Luxor Las Vegas, will be an example of many other cases of using the Egyptian heritage without any compensation, (within the limitations of law and common agreement of the heritage copyright). However, the last ten years many actions were taken but it is still in needed more support and ideas. Two proposals are ongoing to be out of the box; the first, having a competition of how to share with the international community ideas to get the Egyptian heritage rights from those commercial benefits of having Egyptian monuments or replicas. The second is to invent a new theme/brand/science of the Egyptian (Luxor) heritage to be reference to copyright all replicas and such business.

3. Background

Luxor (Thebes for the ancient Greeks); the treasure of world-renowned monuments is one of the major nodes on the Egyptian map of tourism. The form and fabric of Luxor is a story of 5000 years of history and development. This longstanding historical development, accompanied with continuous physical, social and political change, has made it's the conservation of Luxor a rather difficult task. The burgeoning populations combined with the hordes of tourists visiting each year have caused more havoc in the past few decades than thousands of years of erosion. Appeals are made for the safe-guarding of Luxor, studies and planes have been undertaken, but unfortunately, reality has always been frustrating.

On the other hand, a mock-up of Luxor heritage was presented in Luxor Mega hotel project of Las Vegas (1989); the symbol of the American Dream. The Hotel was built as a 30-story pyramid in an Egyptian style, rising from the desert with 4,400 rooms and 120,000 square feet of casino space. Prefabricated "Egyptian" attractions and structures are scattered across the vast floor of the Luxor-replicas of tombs, statues of kings and queens. The business of using replicas to promote "heritage" as a vehicle for tourism is remarkable. Ironically, for many the name "Luxor" today means "Luxor, Las Vegas" rather than the real Luxor of Egypt!!

The purpose of this paper is two-fold: first, to make a comparative study between both the original and the replica in terms of tourism management and heritage conservation, and second, to discuss the notion of heritage reproduction in terms of its copy-right status, and the implications for such right. Specifically, the paper discusses the following arguments:

Tourism industry has different meanings and dimensions in Luxor Egypt as opposite to Luxor Las Vegas

Real culture tourism could include recreation tourism, while recreational tourism utilizes culture tourism.

Tourism management is the art of how to import people from place to another with political, social, economical and technical aspects.

In the time of globalization, discussions should be given to the issue of copy-rights and cultural property (an analogous to "intellectual property") between Luxor Egypt (the original) and its reproduction, Luxor Las Vegas (the replica).

4. Tourism industry and the heritage market

Tourism is a major development form, and one of the world's largest industries and although it is not the only activity that occur in historic cities, its relation to heritage and conservation activities is unique; it is a large export industry and earner of foreign exchange, involves millions of people who spends millions, the largest single item in the world's foreign trade, and is often considered an economic sector with a realistic potential for growth beyond the short term [1]. It also represents a major contribution to national economies.

Defining the tourist industry is difficult. It is not an industry that is grouped into a single heading within the Standard Industrial Classification (SIC). The defining feature of tourism is not the product, but the purchaser, the 'tourist'. Most definitions concentrate on the services that a number of different industries, such as the travel industry; hotels and catering; retailing and entertainment provide to tourists.[2] In the context of foreign tourism,

Fig. 1: Luxor; right: Las Vegas, the entrance; left: Egypt, Luxor temple

Fig. 2: right: 'borrowing' heritage; Dandor temple, metropolitan museum in New York; left: Cloning heritage; Luxor, Las Vegas

the "tourist" has been defined by the United Nations as a "visitor staying at least twenty-four hours in a country other than in which his usual place of residence". Domestic tourism, on the other hand, are visits made within a country by residents of that same country. The purpose behind these visits gives more precise definitions, i.e. business tourism, leisure tourism,..., and culture tourism. It is therefore true that "anything that you can do that creates a destination out of your community is piggy-backing upon a major growing sphere of economics." [3]

For both Luxors' 'culture/heritage' tourism played the main role for attracting visitors. Culture/heritage is a social, political and economic resource. Since heritage is seen as a value in itself, heritage artifacts are suitable for collection, preservation and presentation. National heritage based on national history "explains the distinctiveness of a nation through time", thus it is a valuable tool in increasing national unity and pride, or creating a national image.[4] Furthermore, as an economic resource, culture/heritage is used in various forms. The so-called culture/heritage tourism industry is a "major commercial activity which is based on selling goods and services with a heritage component". Tourism and leisure services obviously play a significant role in this industry, but the manufacturing and sale of heritage products can be considered as similarly important. Aspects of heritage can be used for creating images for places and for promotional purposes. Heritage tourism can be defined as "tourism which is based on heritage, where heritage is the core of the product that is offered, and heritage is the main motivating factor for the consumer"[5]. Cultural/heritage tourism is traveling to experience the places, artifacts and activities that authentically represent the stories and people of the past and present. It includes cultural, historic and natural resources.

Heritage or culture tourism has been one of the earliest forms of tourism e.g. foreign expeditions in Egypt as early as the 19th century. On the other hand it still maintains the largest slice of the tourism industry in Egypt where Luxor is a major attraction. On the other side of the world, tourism is number one, two or three in economic earnings in 47 of the 50 of states in the USA [6]. Although forms and reasons of tourism are quite different from those of Egypt, "heritage" was "borrowed" (e.g. Dandor temple) or even cloned (Luxor, Las Vegas) in order to promote tourism and create what might be labelled as culture tourism. (Fig.2)

It has also to be mentioned that other types of tourism are also of great importance -if not as important as culture tourism (in some cases) in their relation to heritage and conservation activities, in other words "the historic heritage is only one tourism resource among many in the tourist city". [7]

5. Luxor, Egypt; managing antiquity and cultural tourism

Luxor is an obligatory stop for the 'cultural' tourist to Egypt for its incredible wealth of antiquities; the natural beauty of the Theban Mountains and the Nile River, which offer outstanding scenery for all types of tourists; the local village life present in the town and throughout the river valley and it's the mild, dry, sunny weather, particularly in the winter.

Luxor, Thebes was the ancient capital of Egypt, founded at the Nile, where the east bank (sun rise) was for everyday lively activities, the west bank (sun set) was dedicated to funeral functions. Antiquities and monuments dates back to early pharaonic dynasties (3000 B.C) in addition to later Roman, Coptic and Islamic periods. Among others, Luxor's main landmarks include 'world heritage' sites such as: the royal tombs of the Kings' Valley, Queens' Valley and the Tombs of the Nobles. It also includes master pieces such as the Colossi of Memnon, Karnak Temple (the most imposing Pharaonic temple in all of Egypt) and Luxor temple. They represent some of the finest examples of mankind's early civilization and rank among its greatest cultural achievements. It has therefore always fascinated travelers from all over the world.

Tourism in Luxor has been a major economic activity of most of its population, as it is the source of various jobs and business opportunities. The local economy is therefore largely depended on tourism and has therefore been seriously affected with any regional or political disruptions.

This unique cultural heritage continues to attract visitors from all over the world in ever growing numbers. Inconsistently, their dedication to viewing these treasures is becoming a threat. In the tombs, their very presence is becoming detrimental to the quality and preservation of the paintings, and in temples, their increasing number and the lack of any effective crowd management means waiting and jostling, elements that detract from the cultural experience. At the same time, the virtual absence of facilities for other tourist activities means very short stays in the area, lessening the benefits to the local economy, and less flexibility in scheduling visits to the cultural sites.

Thus, the biggest challenge facing tourism in Luxor, Egypt is ensuring that tourism does not destroy the heritage that attracts visitors in the first place. In addition, tourism itself is a competitive, sophisticated, fast-changing industry that requires dedication to keep up with. While tourism is generally considered a clean industry (no smokestacks or dangerous chemicals), it can put demands on community infrastructure, such as roads, airports, water supplies and public services like police and fire protection which, in turn, have adverse effects on heritage sites.

6. Luxor Las Vegas: tourism management and heritage consumption

Las Vegas is a booming metropolis City that is one of the largest fantasy land and fastest growing cities in the world. The increased tourism in Las Vegas has created a market for more tourist accommodations. One of those mega structure projects is "Luxor Las Vegas".

Fig. 3: Luxor; right: Las Vegas; left: Egypt

Fig. 4: Luxor (Avenue of the Sphinxes); right: Las Vegas; left: Egypt

The Luxor was built in 1993 with 300 million dollar budget that insured it to be very impressive. The 30 story Egyptian themed resort had, at the time of its completion, 4,400 guest rooms, 10 river barges on an indoor Nile, and a reproduction of King Tut's Tomb. The casino sizes are full of amazing area of card tables and slot machines.

In a brief context description[- Design philosophy of the architect "Veldon Simpson", is to continue to push the limits of the imagination… "in search of things as they ought to be"… and to clearly understand that in the final architectural statement; the owner's success in the business venture is as important as the Architect's dream…

The Luxor was an urban project, a massive concrete structure of Egyptian wonder, built in glass and concrete pyramid with a large sphinx replica comes into view, which was intended for use by tourists for, accommodations, recreation, and of course gambling[- Veldon Simpson-Architect, Inc. http://www.veldonsimpsonarchitect.com/]. Rings of continuous corridor balconies overlook the atrium, each a higher level overhanging the one below by about 2 m. The entire atrium is the largest atrium-hotel in the world. The elevators were a feat for the engineers because of the massive incline, and moved up at a 39 degree angle. The main entrance is a steel-framed replica of the Sphinx in Egypt. Though the proportions are not exact the Sphinx still plays its role quite well. The Sphinx had to be brought to scale with the pyramid so it is slightly larger than the true Sphinx. In contrast the Las Vegas Luxor pyramid is slightly smaller than The Great Pyramid at Giza, but they were restricted by federal aviation rules, due to the fact that they are a mere 1 mile from the airport. The landscape below the Sphinx was done up to look like a wild Egyptian landscape right off the Nile with the use of authentic plants. While walking along the sidewalk towards the Luxor you will first meet a full scale replica of sphinxes-avenue (rows of ram-headed sphinxes) similar to those found at the Temples of Karnak and Amun at Luxor. (Fig.4)

The projects intention was to create a pleasing and memorable entry experience into the hotel itself. The statues are exact replicas in form and the extensive use of palms in the hotel landscape, not only did they make the land look more like a desert they gave great amount of shade while were exploring. There is also a tall obelisk with the name of Luxor similar to the one in front of Luxor Amun temple in Egypt. (Fig.5)

The rooms are provided with Egyptian-style wardrobes decorated with hieroglyphics, containing a closet, drawers, and a Panasonic television. The headboards sport the cartouche of Cleopatra and there is a replica fresco hanging on the wall. The bathrooms are large in proportion to the rooms. Most people seem to be curious about the shape of the rooms, as if they might be small pyramids. They are rectangular and the window wall slopes inward. The upper floors are accessed via the so-called inclinators. A conventional vertical elevator would provide a short ride unless it was located in the center of the atrium, so at each of the four corners of the Luxor there is a slanted shaft which carries the elevator car up and down at a 39° angle. Each corner inclinator stops at a certain set of floors; person must know which one to use to access a given floor.

Fig. 5: Luxor (entrance obelisk) ; right: Las Vegas; left: Egypt

The casino itself doesn't seem especially large, but this may be due to the circular layout around the central obelisk. Thematic slots such as the "Pharaoh's Treasure," "Golden Cobra," and "Valley of the Kings" are supplemented. As will be seen, the designers of the Luxor apparently felt they could take the Egyptian theme only so far. The casino is quiet compared to most; the ringing bells of paying slots are oddly subdued, perhaps to render the pyramid more authentically tomb-like. The shops are appropriately "Egyptian Sundries" ("for all your Nile needs" including, for the sweet tooth, Gummy Mummies). Restaurants include the Isis, the Sacred Sea Room (tomb relief reproductions of Egyptians fishing), and the Pyramid Cafe.

This Egyptian theme was further explored and utilized with a number of attractions; the "Nile River Tours accompanied by knowledgeable archaeologists", 3D movies of the past and Egyptian objects, and the most worthwhile as well as the most Egyptian-themed attraction at Luxor was the "King Tut's Tomb and Museum" located on the lower floor (not existed now). That was a replica of Tutankhamen's tomb and treasures "just as Howard Carter found it in November 1922." A cassette tape player guides guests past the rooms.

To conclude, the Luxor Las Vegas project has extensively capitalized on LUXOR's heritage to promote the gaming and tourist industry in Las Vegas. It has found a way to bring these remarkable, distant places to those who are unable to travel to see them (10,000 per season). In other words, it has developed "cultural (heritage) tourism industry" in an alien gaming context.

But from the other hand casino gaming is considered a legitimate form of entertainment, and it's not going away. Annual casino revenues are higher than revenues from movies, spectator sports, and theme parks combined. Even so, only about 30% of all American adults have ever been to Las Vegas.

7. Heritage: Authenticity and copyright

Heritage is a broad concept that includes both natural and cultural environments where "something transferred from one generation to another" [8]. Within the built environment it is the part which possesses a special value to the user, and is therefore a major resource in the particular urban environment. In the cultural context, heritage describes both material and immaterial forms, e.g. artifacts, monuments, historical remains, buildings, architecture, philosophy, traditions, celebrations, historic events, distinctive ways of life, literature, folklore or

Fig. 6: Luxor (King Tut's Tomb); right: replica; left: original   Fig. 7: Pyramid; right: Luxor Las Vegas; left: Giza, Egypt

education.

Being mostly a property and a commodity, proprietary rights have to be well defined in order to maintain its survival and prosperity [9]. The nature of these rights varies according to the different categories of heritage as well as the nature of ownership. If certain values are attached to a particular part of the heritage it becomes necessary to have "heritage tenure", i.e. a form of tenure which surmounts the proprietor protected in law. Many question marks arise here; what are these certain values? Who attaches these values to that particular part of the heritage? What kind of tenure-ship is the heritage tenure? Very often Cultural is the common word to describe these values and it is often that part of heritage which is popularly called heritage.

Managing heritage has a primary objective of communicating its significance and need for conservation to its host community and to visitors. Cultural tourism as explained before has therefore been a major vehicle for marketing items of 'cultural' aspects including archeological sites, history, the customs and traditions of people, and way of life. Cultural/heritage is irreplaceable sources of life and inspiration.

Despite the fact that Luxor, Egypt has a universal dimension as for being a part of the 'World Heritage', its Egyptian authenticity is not disputed and it is therefore that International Cultural Tourism Charter (ICTC) [10] states in sec 4.1 that "the needs and wishes of some communities or indigenous peoples to restrict or manage physical, spiritual or intellectual access to certain cultural practices, knowledge, beliefs, activities, artifacts or sites should be respected", the charter also states in sec. 4.2 that "The rights and interests of the host community, at regional and local levels, property owners and relevant indigenous peoples who may exercise traditional rights or responsibilities over their own land and its significant sites, should be respected. They should be involved in establishing goals, strategies, policies and protocols for the identification, conservation, management, presentation and interpretation of their heritage resources, cultural practices and contemporary cultural expressions, in the tourism context."

On the other hand, considering these antiquities as artwork, they are likely to have copy rights where "only the person who created the copyrighted work is legally permitted to reproduce, perform or display it, distribute copies of it, or create variations of it; any unauthorized exercise of any of these rights is called "copyright infringement" and is actionable in federal court.[11]

This notion can be further emphasized by the fact that in some legislation such as the Greek law it is clearly stated that "The use of any element of Greek cultural heritage, as a trade-mark or pattern or sample, is strictly prohibited without the prior permission of the Hellenic Ministry of Culture. [12]

Furthermore, such project (Luxor Las Vegas) cannot be exempted as "fair use" [13] as it is: First: utilized for pure commercial activities: i.e. tourism, recreational and entertainment where the financial value has been the main objective, while depriving Egypt from possible tourism as by providing a closer replica. Second: the amount reproduced has been

Fig. 8: Luxor nights; right: Las Vegas gaming; left: Egypt sound and light

significant; the whole character of Luxor Las Vegas is from the Egyptian theme. Not only the Architecture style but also every single item related to tourism and gaming business. (The interior, the machines, and the services even the coins used for gaming). And finally, the negative effect of reproducing Luxor Egypt, upon the potential market for original Luxor as for many it means "Luxor Las Vegas" rather than "Luxor, Egypt".

8. Tutankhamen's replica and the case of competition

Implicitly there were recently several events as a reason to reopen the case of replica and copyright. A group of Spanish, British and Swiss companies built on November 2012 a replica model of Tutankhamen's tomb by a cost more than 40 million Egyptian pounds, and would like to give it as a gift to Egypt the question where to find a place to be put it in Egypt (if it will be in Luxor, Harghada or Sharm El Sheikh), finally the decision was Luxor. The consequences of Egypt accepting such replica as a gift means that the same company could have the rights to make another models, and then Egypt couldn't demand its rights to sell or trade.

On the 30th 2014 April the facsimile of the tomb of Tutankhamen was opened to the public by two ministers, the Governor of Luxor, the EU ambassador and about 25 other ambassadors from the different countries. A facsimile has been made that is identical to the original at normal viewing distances. It has been placed within a small museum that reveals why it looks as it does and why it is so difficult to preserve something that was built to last for eternity but not to be visited. On that day the announcement of the copyright competition was to balance the debate of that event. Is such project will conserve the original tomb from all the tourism and visit problems or this replica will open the market for similar projects and the original will lose its unique. And to what extend the business versus the heritage copyrights.

Considered as conviction more than propaganda, whether all those replicas treasures of Egyptian civilization, why in present couldn't keep the original. The issues were the beginnings of this case represented by the "Luxor - Las Vegas, compared to Luxor - Egypt" or the "land of the pharaohs in Dubai land" and other projects in China, Japan and America," work and study" as well as demanding the rights of intellectual property, which referred to. For example, a simple comparison between Luxor and successes replicas Luxor (which visited by 50 million tourists a year) and the suffering Luxor the original (we can how many tourists in Egypt, not in Luxor) it shows us the importance of looking at intellectual property rights with regard our civilized heritage.

And because there is a place to argument the issue of our rights must be in the priorities of this country for those who gain from the issue of architectural heritage, and by the puzzled and reference issue between being an issue of Egyptian Antiquities, Or the issue always raise as an architect, I emphasize the Egyptian monuments were ideas and Architectural work and by time transferred to became antiquities.

9. Civilization is generated from a womb of a civilization

On 21st May 2014, under supervision of Luxor government and preparation of Center for Sustainability and Future Studies British University in Egypt, it has been announced and invited to Luxor Olympiad through an international competition for Luxor, Egypt and the world youth to enable youth, scientific research, Luxor civilization to awareness return with the end of Luxorians Egyptian and the whole world illiteracy, by our treasure and chances and enormous possibilities and a waiting future to our condensations and preparing to certain steps for coming generations.

That is Egypt and Luxor deserve, we need a model in Egypt to make a vision to immigrate to a scientific and practical stages to make a state policy, planned stages and programmed sectors compatible with our possibilities, at last particular projects. Luxor why? It's a minimized clear model for Egypt, 7000 year civilization and enormous treasures and possibilities, in counter the man of Luxor suffering from quality type of life in every side housing, education and transportation etc, that deserve the capital of the first human civilization, and the international tourism inheritance, and deal with responsibility and work necessity.

Hours before the announcement of the competition and I'm at the stage of preparing I remembered the events throughout the last 30 years has been associated with Luxor in studies of MA and Ph.D. and multiple research full of ambition , optimism, sweat, effort, the opportunities and attempts, and with it was frustration and despair, so there were internal multiple questions before competition announcement from the heart to the mind, between despair and hope for why we declare today, will it be successful round or missed opportunity in the short lifespan.

What originally competitions do it is a way or a goal, and what is the management mechanism of the event? Is it professionally, random or fatalism? What does Luxor mean for Egyptians - Nubian and Sinai Bedouin and the people of the valley in different locations and backgrounds, and for the world - Japan, Australia, America, India, Italy, Mexico or Mozambique?

The word of Challenge the less word described the stage and challenge or invention required to talk in less than half an hour about why and how and when.

The press conference was attended by the people of Luxor, businessmen, banks, companies the media and youth and activities began with an address of the governor (which is rarely found enthusiastic of those ideas to involve young people and Scientific Research) and the governor stress on the goals of competition to reach the sustainable development of Luxor.

The competition: "Branding Luxor and intellectual property rights", Luxor is layers of successive history, and the capital of human civilization and it has heritage, distinct monuments and every site deserves to be a marketing registered sign, the local community of each area linked with industry production and emphasizes on the subjectivity and value. Luxor is: A set of heritage environments of temples of Karnak, Luxor, Al Kebash road, the tombs of the Valley of the Kings … etc , which can be more than 15 heritage environments have the archaeological heritage and urban to be as a shrine with its own program and registered date mark and put it in imagine logo and features quality brand marketing product for specific cultural and entertainment to each environment within the tourism program for the city of Luxor. Integrates with the goal of marketing heritage environments Luxor the second part of the competition was to discuss the global intellectual property rights and reproduction of Egyptian heritage in profitable projects spread across the world.

10. Summary and conclusion

The issue of intellectual property rights of urban and architectural heritage can be summarized in points:

The intellectual property rights have been put to protect the innovation and creativity in all different fields whatever It also includes food, clothing, culture and music.

Thus, who gain money from advertising and use our environment and cultural, historical and architectural heritage doing copies; we have intellectual property rights must pay and must ask for them. The profits from cloning of that culture and historical inheritance we can at least direct it towards the development of this heritage in order to maintain it through the detection and preserve monuments and restored instead of waiting for the directed grants and donations that we cannot even control in their plans.

it's more than just rights of cloned heritage but includes also the original Egyptian heritage which shown in the global museums (such as the head of Nefertiti) that we not able to get it back). the head of Nefertiti and many other unique monuments considered as a benefit for its visitors, only way for Egypt is getting the proportion of the yield as a right of intellectual property rights (at London's museum many models as a souvenir or gift in the bazaars and museums, such as the Rosetta Stone the eraser sold by 2 pounds, and the paper sizes different from 6-15 pounds and other examples need folders).

That all of Greece and Mexico and other countries which have Old cultural heritage keep its property in order to maintain the intellectual property rights of their heritage.

There is no doubt the political and economic dimension, the reality and falsification of the concepts in the memory of the global community, we are aware that the human and architectural heritage is the meaning of historical events are based on conflicts perhaps occur after hundreds of years.

These cloned projects raise questions include a lot of dimensions: such as (the ethical dimension) is it possible to use the cultural Egyptian heritage in immoral acts such as gambling by the name of tourism? And (marketing side) all products, games, and even the currency used inspired of the Pharaonic heritage, will be manufactured and trade for importation from Egypt and what's the benefits for Egypt, or China factories is the beneficiary? (Side of our intellectual property rights: When searching for a word Luxor in the internet you will find that most of the sites are talking about Luxor - Las Vegas not Luxor of Egypt.

Egypt could to earn millions by sharing continents, if the Legal entities may take time to prepare, we could create methods in order to participate all of us in it depending on the appeal of Egyptian citizen to conscience of the universal citizen principle of if I have studied - I read - I loved civilization of Egypt This civilization sweep away layers of stealing, neglect and the human overcrowding. The international community may impose on the companies and institutions that profit from the Egyptian heritage a percentage of their profits to maintain the heritage of Egypt and construct its future.

Tourism in Las Vegas and Luxor is the main activity of their economic development, where it is clear that Las Vegas is doing much better terms of tourist numbers and financial revenues, this paper does not aim merely to blame the Luxor project for this. Yet, this can be directly referred to the obvious reasons of political and economic stability in both regions, and we also acknowledge that While the Luxor project is a mere economic development run by a private business company, Tourism in Luxor city is part of a community development process where a complex of social, economic, political and cultural factors are involved. This paper finally aims to highlight the following:

Luxor city should – as a copyright holder- financially benefit from all replicas such as the Las Vegas project unless it is acknowledged as a universal property of the international community which then have to hold full financial responsibility of the management of this ' universal' heritage and not only the moral and technical support of the 'World heritage' convention.

References

[1] Williams, A.M. & Shaw,G. eds, "Tourism and economic development", Belhaven Press, London, 1988. p 1.

[2] Eurostat, Community methodology on tourism statistics, 1998

[3] McNulty, R., "Lessons from North America", International conference; Heritage and successful town regeneration, Council of Europe, Halifax, U.K., 24-27 Oct. 1988, Strasburg, pp.21-25.

[4] Tunbridge, J.E & G.J. Ashworth, "Dissonant Heritage. The Management of the Past as a Resource in Conflict"; John Wiley & Sons, Chichester, 1996.

[5] Swarbrooke, J. , "The Future of the Past", Heritage Tourism into the 21st Century; In: A.V. Seaton ed.: Tourism. The State of the Art; John Wiley & Sons Ltd, Chichester, 1994, p.222-229

[6] McNulty, R. (1989) p.2

[7] Ashworth G.J. & Tunbridge, J.E, "The tourist-historic city", Belhaven Press, London.,1990.p.51

[8] Veldon Simpson-Architect, Inc. http://www.veldonsimpsonarchitect.com/

[9] Nuryanti, W., "Heritage and Postmodern Tourism"; Annals of Tourism Research 23(2), 1996, :249-260

[10] Liechfield, N. "Economics in urban conservation", Cambridge University Press, Cambridge, 1988.

[11] INTERNATIONAL CULTURAL TOURISM CHARTER, " Managing Tourism at Places of Heritage Significance" , 1999. http://

www.international.icomos.org/charters/tourism_e.htm

[12] Wilson, Lee. "Copyrights, Trademarks, Patents and the Graphic Designer", Communication Arts. Coyne and Blanchard, Inc. Palo Alto, CA. Dec. 1994:158.

[13] Official journal of the state republic of Greece, "Copyright related rights and cultural matters, law 2121/1993", First issue, issue number 25, 4th March, 1993, http://www.culture.gr/6/64/law2121.html

郑军德
Zheng Junde

浙江师范大学美术学院副院长、教授、硕士研究生导师,中国建筑学会会员,浙江省美术家协会会员,浙江省水彩画家协会会员,浙江省流行色协会理事,浙江师范大学社科联文化艺术服务协会会长。

研究领域:校园环境文化设计及美丽乡村环境设计。

近年来主持完成各级各类课题(纵问及横向)120多项,出版专著及教材3部,多篇(件)论文及作品在《文艺争鸣》《当代文坛》等国内核心刊物上发表。艺术作品多次在省级及以上各种展览中入选并获奖。

Vice Dean, professor, supervisor of postgraduate of the School of Fine Arts in Zhejiang Normal University; member of the Architectural Society of China, Zhejiang Artists Association and Zhejiang Watercolor Artists Association, Council Member of Zhejiang Fashion Color Association; President of the Art & Cultural Services Association of Zhejiang Normal University Social Sciences Union.

Research field: campus environmental culture design and beautiful countryside environment design.

In the past few years, he presided over more than 120 tasks and projects at all levels, published three monographs and textbooks, many articles, papers and works are published in domestic core publications like "Art Contention", "Modern Literary Magazine". His artistic works have been included in various provincial or high-level exhibitions and won many awards.

新型城镇化背景下浙江小城镇人居环境营造之探索

郑军德、徐艺文

【摘要】介于城乡之间小城镇的发展在新型城镇化发展特殊阶段有着它独特的意义和地位。而城镇化的实质不仅是实体空间的城镇化,更是人的城镇化,是人居环境的城镇化。因此,本文对浙江小城镇发展过程中人居环境的现状、营造原则及营造方法进行一定的分析阐述,希望能以此引导和促进我国小城镇人居环境朝着科学、可持续的方向发展。

[Abstract] The development of small towns between urban and rural areas has unique significance and status at the special stage of new-type urbanization development, while the essence of urbanization is not only the urbanization of physical space, but also the urbanization of human and human settlement. Therefore, we analyze and expound the current situation, building principle and building methods of human settlement in the development process of small towns in Zhejiang in this article so as to guide and promote the scientific and sustainable development of human settlement of small towns in our county.

　　回顾建国60多年来,我国城镇化水平由1949年的10.6%上升到1999年的30.9%。到2013年年底,我国城镇化水平已超过54%,是前14年我国城镇化速度的1.8倍。党的十八大报告提出,要坚持走"中国特色新型城镇化"道路,其中不仅增加了城镇化的相关内容,提升了城镇化的地位和作用,更明确了要结合中国国情和各地实际情况,分类引导,形成多元的大中小城市和小城镇协调发展的新型城镇化道路。

　　而长期以来我国研究领域将较多的目光投向了城镇化体系的两头——大城市和农村,却忽略了城镇体系的基础——小城镇在城镇化建设中的积极作用,忽略了城乡一体化的桥梁——小城镇中人居环境营造的重要性。这里的小城镇人居环境主要是指在小城镇中围绕人这个主体而存在的一定空间内的构成主体生存和发展条件的各种物质性和非物质性因素的总和。

　　浙江,一个没有特大城市的省份,能够成为中国城市化水平最高的地区之一,小城镇发展是功不可没的。而浙江的小城镇发展又以地方经济的快速发展为基础,这也使浙江小城镇的经济快速发展和人居环境相对滞后的矛盾日益增强。

一、浙江小城镇人居环境现状

　　改革开放后,浙江一些农民开始从田间地头走出来去务工、经商、办企业,并逐渐形成了一批人口产业集聚程度较高、经济实力强、发展动力足、产业特色鲜明的小城镇。作为城乡之间、工农之间人口、产业、服务等市场要素的重要聚集地的小城镇,在快速发展的过程中普遍存在人居环境发展不协调的问题。

　　(一)空间缺乏科学规划、合理布局

　　浙江小城镇在经济快速发展和人口大量涌入的情况下,常常忽视了小城镇空间的科学规划和布局,忽视了空间布局与空间优势和特点相结合,忽视了空间布局与空间中人力、文化、自然环境的关系,也忽视空间布局与节约资源和保护环境的关系。使小城镇空间在功能定位、区域划分、资源使用等方面都未能达到协调统筹发展,使城镇发展粗放,资源大跨度调运,极大增加了经济社会运行和发展的成本。

　　(二)忽视与周边区域的协调发展

浙江小城镇作为连接城乡的重要纽带，在城乡发展结构中具有承上启下的作用，既是工业化的重要载体，又是农业产业化的服务依托，因此，小城镇的发展必不能脱离周边区域独立发展。但在对浙江小城镇现状分析过程中，不难看出大多数小城镇建设过程中对周边区域的发展格局、发展结构和发展特色都没有作深入的调研和分析，基本从自身考虑、从主观出发来制定小城镇的发展规划，这样必然忽视其与周边区域之间的空间、经济、社会关系等方面的同步和协调发展，使小城镇的发展缺少可持续性。

（三）工业污染加剧，生态环境破坏严重

随着经济的快速发展，小城镇也不得不为巨大的生态环境破坏而买单。过去，村镇的废弃物基本上都是有机物，经过生化处理后作为肥料基本不会对环境造成污染。但是随着小城镇人口的增加、工商业的发展、生活水平的提高、农业新技术的应用等，造成环境负荷高，排污强度大，城镇污染情况严重，尤其是大气环境质量严重受影响。浙江地区小城镇工业粗放型发展、分散式的布局和薄弱的管理体制也是污染得不到及时治理的主要原因。

（四）传统特色与个性逐渐丧失

浙江地区优越的自然地理条件和悠久的社会发展历史，形成了独特的地方建筑风貌和江南小镇特色。在经济迅猛发展的同时，人们传统的生活方式和小城镇建设模式发生了改变，造成传统建筑空间模式在不断面临着巨大的冲击。一方面，小城镇的传统文化和传统建筑没有得到很好的保护，另一方面，小城镇在规划和新建的过程中没能结合当地传统特色，不注重原有的地形地貌、景观特色，而盲目追求与小城镇尺度不相符合的宽大马路，高大厂房、时髦别墅等，导致小城镇之间的结构形态雷同，原有的特色和个性逐渐消失。

（五）忽视以人为本的小城镇人居环境营造

目前小城镇人居环境营造更多是从单一的建筑、景观等作为出发点，注重物质层面的实体要素，而忽视了以人为本的人居环境营造所要提供与城镇经济发展水平相适宜的基础设施和基本公共服务，还要考虑到居住人群的生活情趣，信息交流与沟通，人们的安全感与归属感，等。《马丘比丘宪章》中提及"深信人的相互作用与交往是城市存在的基本依据"。大多数小城镇人居环境营造中缺乏对生态环境、人文环境等人居环境组成因素的综合考虑，而是人为地将区域、镇区、乡村、建筑等从整体中分离出来，因此，设计间缺乏逻辑关系。[1]

二、新型城镇化背景下浙江小城镇人居环境营造的原则

在新型城镇化建设过程中，注重小城镇建设与人居环境的协调发展。对经济、环境、社会协调发展以及区域整体协调发展是浙江小城镇人居环境营造的必然趋势，小城镇的发展必须是可持续的、稳定的。因此新型城镇化背景下浙江小城镇人居环境营造应当遵循以下原则：

（一）区域统筹协调的原则

区域内的统筹协调，即指城乡之间的联系和协作，统筹配置资源取得整体最佳。在浙江，以经济发展为基础的小城镇发展成为我国最具代表性的城镇化发展模式之一。在其经过初期的经济自发性和粗放型发展之后，要更好地营造小城镇人居环境就必须实现城乡统筹和协调的原则，即对区域内生态环境、资源能源、产业发展、社会文化、科技创新以及公共政策等方面作整合总体安排，以达到空间布局及社会发展的可持续性。

（二）生态绿色和环境友好的原则

在小城镇的人居环境营造当中，需树立环境的可持续发展观，克服营造过程中牺牲生态环境加快发展的

短期行为，将人工环境和自然环境有机结合，遏制大面积硬质地面，对人工建筑不断增加，绿化植被面积锐减，水体和空气严重污染，过多地耗用能源、水和土地资源等一系列问题作综合考虑。以生态环境、资源保护为基础，赋予小城镇更多的生机和美感，更好地吸引投资和拉动消费需求和经济增长，从而营造人口规模适度、经济快速发展、布局规划合理、生态绿色和环境友好的人居环境。

（三）保护地方特色、传承历史文脉的原则

浙江省数量众多的小城镇，各有经济、文化、民俗、风情等方面的地域差异，因此，在加速城镇化进程营造人居环境的过程中，只有注意保护地方特色，传承历史文脉，富有特色的小城镇才具活力和凝聚力。如浙江小城镇各有鲜明的产业特色、独特的地域特色、鲜活的生态特色、惇厚的历史文化特色，而这些都是小城镇在人居环境营造过程中必须关注和保留的。新型小城镇人居环境营造不仅仅是客观物质上的建设，而要更深一层去"做文化"，力求做出特色、做出品牌。我们必须创造能被历史见证的小城镇，必须保护有价值的建筑和环境，遵循地方传统和文脉的原则。[2]

（四）功能性与宜居性相结合的原则

新型小城镇人居环境营造就是"以人为本""为人服务"，其功能性是非常重要的。一方面要考虑城镇交通流线的合理布置与通达性；另一方面也应考虑到生活设施、公共设施及景观绿化等因素所构成的整体美感与协调性。在小城镇人居环境营造过程中，除了强调功能上的完善与便利，也应该注重小城镇与乡村生态环境的和谐发展，并在此基础上达到居民对生产生活环境的舒适、美观等要求。

三、新型城镇化背景下浙江小城镇人居环境营造的策略

《老子》云"安其居，乐其业"。良好的人居环境不仅是社会建设的需求，更是人们安心生活的需求。随着小城镇经济建设的发展和人民生活水平的提高，人们已从对温饱的需求上升到对高品质生活的追求，而创建良好的人居环境则是这一追求的根本。小城镇人居环境营造需要树立全局观念，分层次、分系统地剖析当地人居环境，再面对实际进行研究和解决。

（一）政府主导，推进人居环境合理布局

1. 产业结构空间合理布局

小城镇中各产业的空间合理布局是实现小城镇人居环境可持续发展的关键。工业区区位的选择要从区域总体发展战略的角度出发，除了考虑影响生产、经济的因素之外，还要把人、环境、空间等整体区域内各项协调统筹发展，不能顾此失彼；为了实现现代农业的发展，农业耕地红线是必须要坚守住的底线；作为第三产业，基础型的公共设施建设对改善小城镇面貌、优化小城镇建设环境具有巨大的潜力。

2. 环境空间的系统布局

当下小城镇营造中较为凸显的一个问题就是对于空间环境缺乏系统的规划，在人居环境营造中并没有将"建筑—地景—城镇规划"三者进行系统的分析设计，需结合镇域内的自然条件和发展趋势进行功能区域规划、道路交通规划、环境景观规划等，逐步建立多层次、多类型的开敞空间，形成系统。

3. 生活空间的宜居布局

对生活空间的布局要以宜居性为原则，顺应自然、因地制宜应该是小城镇空间格局规划的一个主导思想，规划和谐自然的小城镇空间格局，建设环境清洁、优美、安静、舒适的生活空间是人居环境建设的重要部分。

（二）全民参与，加强人居环境共建意识

全民参与是人居环境建设的要求，更是人居环境可持续发展的必要条件。联合国环境与发展大会在《21世纪议程》第23章第2节明确提出"要实现可持续的发展，基本的先决条件之一是公众广泛参与决策"。

1. 树立人居环境意识

走可持续发展之路，保护和改善生态环境，促进人居环境可持续发展，迫切需要社会民众对当地人居环境的基本情况有所了解，明白人居环境危机更多的是由人为因素造成的人居环境恶化，并树立一种危机意识。培养公民自觉维护人居环境的道德意识，从而在全社会形成一种关注、改善、保护人居环境的良好社会风气和社会规范，并形成有利于人居环境营造的和谐社会规范、树立人居环境伦理观。

2. 加强全民参与意识

小城镇的人居环境营造是以解决百姓生活的实际问题为基础的，是以提供完善的城镇基本功能，改善城镇环境质量，提高城镇宜居度为目的的，因此让小城镇居民一起加入到和谐的人居环境的营造中来是必需的。只有政府主导，全民参与，沟通共建才能共同营造小城镇的良好人居环境。全民参与人居环境建设行动，除参与政府决策外还可以加入相关社会组织。这里主要如各类环保组织、社区居民委员会等。而组织性、私有性、非营利性、自治性及志愿性是这些组织的特殊属性，它不仅能弥补市场与政府的一些缺陷，而且在人居环境营造领域中发挥出重要的治理和监督作用。因此，要合理的利用民间非正式组织的力量，搞好小城镇人居环境。

（三）数字量化，建立发展人居环境长效机制

人居环境的营造需要根据实际情况进行量化比较，制定和完善适合本地人居环境的建设指标体系、评价体系和管理体制，只有这样才能更好地指导全民实施操作，实现资源的优化配置、节约和保护。如制订有关环境保护、资源管理、生态环境建设、基础设施建设和维护、公共服务设施的建设和维护、社会保障等方面的法律法规制度，用法律法规制度推动小城镇人居环境的健康发展。

（四）逐个突破，创建人居环境和谐发展

小城镇的人居环境介于城市和乡村两种人居环境之间，它一方面具有乡村更接近自然生态环境的特点，另一方面也拥有城市性的相对完善的公共和基础设施。但因小城镇近几年的快速发展，生态破坏、环境污染、居住条件不平衡、基础设施和公共服务滞后等都是困扰小城镇人居环境发展的重要原因。因此，小城镇要得到人居环境全面和谐发展，必须在考虑全局的基础上抓重点、难点逐个突破。

1. 加强生态环境营造

生态环境保护是小城镇人居环境可持续发展的重要组成部分之一。小城镇生态环境保护必须首先以区域环境保护战略为宏观指导，对农业生态绿地、自然湿地、水源地和森林等生态敏感地区划定整体保护范围，在此基础上，因地制宜地建立小城镇内部生态系统，如增加镇区公共绿地，建设以生态公园为主的城镇绿心，增建以块状绿地、带状绿地为主的小游园和小绿地，强化道路绿化和滨水绿地的建设，等。其次还要加大景观综合整治力度，改变原有环境中的"脏、乱、差"现象，使环境和景观面貌得到明显改善。

2. 完善基础设施和公共服务建设

小城镇的生活基础设施的建设一直滞后于经济的发展，新建的基础设施项目进展缓慢，原有的基础设施又缺乏必要的维护，造成城镇基础设施供给不足的局面。[3] 小城镇基础设施区域统筹规划及其优化配置与共享是城镇规划建设重点。市政基础设施上，排水系统、交通系统等应适当地加大规模，来满足居民们的生活

需求；在基础教育上，重点建设镇区中心学校，提高教学质量；在社会福利方面，完善社会医疗保障制度，扩大养老院规模。适度发展满足较高层次需求的公共设施如影剧院、文化活动中心、广场等公共建筑，对居民以寓教于乐的方式来引入现代生活观念。

3. 创建人居环境良好的生态社区

社区环境建设是人居环境建设中的重要部分。着力营造绿色生活和行为方式，创建有安全感、归属感和荣誉感的生态社区，以点带面地推进小城镇的人居环境建设。通过生态社区的建设实现资源能源利用最大化，倡导社区和小城镇的共同发展。

结语

新型城镇化背景下的浙江小城镇人居环境的营造中，应充分合理地设计镇域空间布局，保护生态自然，尊重地域特色，因地制宜，设计满足居民需求的宜居宜业的可持续人居环境小城镇。优秀的小城镇人居环境应该能随着时间的推移不断变化，是立足于可持续发展战略、凸显小城镇特色的人居环境。通过人居环境营造可推动小城镇生活生产的发展，从而使传统的小城镇人居环境迸发出更旺盛的生命力，实现可持续发展。

参考文献

[1] 吴良镛. 人居环境科学导论[M]. 北京: 中国建筑工业出版社,2001.《城镇人居环境评估指标体系》；陈翀，阳建强，刘源. 我国小城镇规划设计中存在问题剖析——大连经济技术 开发区海滨区域设计研究[J]. 新建筑，2005（3）

[2] 叶耀先.中国小城镇人居环境建设[J]. 中国人口资源与环境，2004（2）.

[3] 张峰.小城镇人居环境建设亟待全面提升[J]. 建设科技，2004（11）.

法比欧·匹斯科普
Fabio Piscopo

艺术家，1950年出生于佛罗伦萨，1975年毕业于佛罗伦萨美术学院，多年来对于研究和试验的无穷渴望促使他不断进行新的艺术尝试：如蜡画，湿壁画，陶瓷画，嵌板画，浅浮雕，青铜。这些尝试获得了积极效果,得到了舆论的一致好评。

他曾在中东，北美（尤其是美洲原住民保护区）生活过多年，近年多在中国访问，与多种不同文化的紧密接触给他深刻而强烈的冲击。这些丰富的体验督促他始终在寻找"前卫"的艺术技巧。

Artist, born in Florence in 1950.He studied at the Art High School and then at the Fine Arts Academy where he graduated with top marks in 1975. The deep need to search and experiment drove him for many years to try encausto, affrescos, ceramics and wood panels, (including burnt ones) bas-reliefs in glazed refrattario, bronze with sculptures, panels and bassorilievos,with really positive feedback. He has spent long periods in the Middle East and in the U.S.A. (particularly in the Native American Reservations) and recently in China , he was in touch with cultures, which gave him suggestive and intense stimulation. These experiences became a drive to search for "Avant Guarde" art techniques.

《修复指南》在建筑遗产保护中的意义及相关研究
——以意大利古镇切尔瓦拉·迪·罗马为例

法比欧·匹斯科普

  切尔瓦拉镇（Cervara di Roma）位于罗马省东北部的山区，距罗马大约100千米，属于拉齐奥大区。切尔瓦拉的名字来源于意大利语的cervo，意思是一只鹿，因为旧时山上有很多鹿故得此名。切尔瓦拉全称是切尔瓦拉·迪·罗马，被认为是西姆布瑞尼山国家自然公园的大门。这个美丽的公园林木茂盛，自然环境优越，是野生动物的天堂。其中的阿涅内河和河谷更是景色秀美的度假胜地。

  从19世纪开始，就有国际艺术家为了寻找灵感来到切尔瓦拉。从目前保存的艺术作品及镇里的美丽传说来分析，许多艺术家来自巴黎，甚至摩尔斯电码的创立者萨缪尔·摩尔斯也曾在镇里居住。上世纪后中后期，佛罗伦萨美术学院的学生在山上石灰岩壁上创作了很大的装饰性雕刻，出现了"艺术家阶梯"和"和平阶梯"两个著名的艺术作品聚集区。

  切尔瓦拉镇的历史文化遗产非常丰富，山顶有中世纪的堡垒遗址，镇中心是翁贝托一世广场。宗教对切尔瓦拉的影响深远，镇上有许多教堂，最大的是"圣母访亲教堂"。每年8月15日的鞠躬圣母游行吸引着无数外出谋生的居民回到镇里欢度节日。当地的美食在节日里更是成为传统的重要组成部分。

  切尔瓦拉的布局是整个苏比亚科地区前现代建筑类型最显著的样本，在历史上同一片区域几百年来都处于苏比亚克修道院的封建管辖下。《切尔瓦拉镇修复指南》一书在罗马出版后受到热烈的欢迎，这也为接下来的工作树立了榜样，比如在罗马的卡斯泰洛城（1990）和巴勒莫的古城市中心（1997），它可以极大地促进修缮工作的进展。有人会问赋予切尔瓦拉镇如此科学而讲究的维护的原因，其中被引用的实例让它看起来更适合一个伟大城市的名望。

  这个问题的所有答案都在罗马行省这个古老山镇里，古镇的中心如今完整地保留着它的特色和前现代的建筑风格，是整个苏拉迪克地区中最珍贵的样本。因此，对切尔瓦拉镇历史建筑的研究表现了修缮维护策略的传播和对阿涅内河河谷的历史的发掘。

  对这种幸运的保护起重要作用的历史和文化环境因素主要归功于苏比亚科修道院，而切尔瓦拉在其中起着的重要作用。1000年前的堡垒，最大限度地利用了防守的潜力，使石墙变得坚不可摧。堡垒位于岩石峭壁上，是整个苏比亚克修道院的主要军事警备区。在接下来几个世纪的城堡的不断发展，使切尔瓦拉成为整个巴迪亚区域内最重要的村镇，在重要性与规模上仅仅次于苏比亚科。

  切尔瓦拉镇的房子并不是非同寻常地广阔且分布均匀，而是认真地遵循石墙结构（从进口处有从苏比亚科的山上采取的石头，有来自本地区的各种自然角硕岩），装饰精致的门与窗，广泛使用的木质工艺品（尤其是门和楼板），具有丰富多样色彩的外部结构，这为多层次的社会历史提供了最有说服力的证明，其中的历史事件已充分体现了这个历史时期的建筑特点、类型以及分布。

  建筑作为一个感性的物体，记录着社会变化和经济现象。新的拥有者对低成本购买的建筑进行重新的修复与整理，与当地传统不相关的普通工业材料的使用进而造成的破坏是司空见惯的。镇上古老建筑物合法的继承人，如果能够通过有效途径，进行充分的准备并且有公共机构的协助，通过好的修复指南的指导，形成新的社会动力，让人们看到他们积极而有效地从事修复保护工作，是与传统相适宜且符合集体利益的。

  和谐地融合了建筑与周边自然环境，诠释这种非凡的联系，建筑语言的简洁，各部分彼此联系的众多空间和系统，都表现了切尔瓦拉的特殊性。《切尔瓦拉镇修复指南》包含完整的有关当地传统建筑的解决方法与形式，对细节的真实再现和建设实施方式的描述，为工人和房屋所有者提供准确而容易的入门指导，可以为没有修复实践经验的人使用。修复指南的意义如下：

一、《修复指南》是修复活动操作的策略工具

在《切尔瓦拉镇修复指南》中,你将会看到第一次研究古老建筑各组成部分的全部目录,据此,既可让从事修复工作的技术人员和工人,也可让普通人以及这笔财富真正的拥有者都能够得到即时且直接的技能操作指南。因此承担职责的修复配套企业,在考虑保温的时候,除了了解有用的内部结构,也能开展他们的日常修复活动。

从某种意义来说,《修复指南》能提供所有的操作方法来进行恢复、修补及重建古老建筑的任何组成部分,无论是墙壁、楼板还是一扇大门。通过对每一个组成部分和执行方式的描述,从而带来完整和真实的重现,给修复人员与建筑拥有者一个有关古老建筑的比较正确的指导,易于修复者入门与查找,并且提供更加实用的使用操作方法。

《修复指南》中,那些有关历史建筑组成部分的章节内容设置,专门把现场调查类型与传统城市规划方法并列,通过即时而又直接的操作技能介绍和单个元素的集合保存,有可能比实际的数据更有条理,从而让拥有稀少成果仅确立了抽象的及教条主义般不思变化的建筑机构,得到有关类型学的专门分析以减少风险的存在。正如权威专家保罗·马可尼所言,古代技术知识的发挥,恢复特有的组成元素和复原历史建筑功能的作用,可类推出可能的结果,这是合理的。如果这是事实,将继续有利于城镇规模的重建与建筑布局中缺失或消失甚至严重损毁部分的复原,也多亏对建筑组成部分持续的更换与静态平衡的改善,使对古老建筑的保护成为完全的可能。

二、《修复指南》是修复主体相关的知识载体

凡是用挑剔的眼光观察前现代建筑的人都会被建筑所存在的细节吸引,但它们十分地复杂。在绝大多数情况下,多由中低水平的工人来实现完成,当然,这也归功于在过去广泛流行的解决方法,即运用图形和模板的目录。这种做法流行于中世纪的建筑师傅群体中,这已获得最终的验证。关于形象艺术(其草图旨在实现壁画从一个大区流传到另一个大区的循环),它一直被运用延传到几十年前。每一个好的工匠,比如画匠和石匠,都拥有铁质模板的真实图集和各种类型作品的实例(如别墅的栅栏,葬礼祭台的三柱门,床头板,栏杆,装饰用品,等)。尽管在现代的批判范围内,但这种做法保留了具有更令人信服的外观,并具有推广的意义,虽然这在19世纪受到许多谩骂,但毫无疑问,它使每个确定的工作类别都有了类型化的图示。同样,每个装饰品都在稳定的常规基础上建立了准确的传播目录,从中可寻得根源。

《修复指南》试图在吸取精华的同时,尝试复兴这个古老方法中积极的方面,以此来为修复计划提供一个真正合适的指导,通过收集图形的方法,最终的目的不仅是为了禁止篡改,更主要是为了打击近年来的投机性破坏,达到保护城镇的目的。需要说明的是,假如修复在本质上只是维修,那么服从规则是不错的选择,若根据更复杂的标准,要设计独特的原作及对特殊案例的校准,它必须保持设计者的自由介入权,那么,这其中便需要注意元素的复杂性以及需要考虑到方案渗入的问题。

三、保留切尔瓦拉老城区的建筑整体性

切尔瓦拉镇的建筑没有出现任何特别的建筑方面的紧急情况,古老的堡垒在几个世纪已完全毁灭,现在甚至难以辨认。访亲教堂是切尔瓦拉最大和最主要的教堂,也没有任何原始隐喻或者显著的特征。尽管如

19世纪的切尔瓦拉镇　　　　　　　　　　　　　　　　现在的切尔瓦拉镇

此，切尔瓦拉体现了建筑与自然环境的联系，超凡而极度和谐，展示了很大的交际性功能。该城镇的建筑语言是极为简洁的，几乎没有任何隐藏含义蕴涵在空间里，空间和建筑双方形成了相互定义的关系。这种平衡的关系当然存在着极端的脆弱性。这种脆弱性来自颜料的质量，建筑的材料，但是值得不惜一切代价保留，要严肃地处置材料内部被危险地侵入的成分以及相比以前有所新增的技术与结构的改变。诱惑总是存在，建筑拥有者会追随消费主义和短暂的潮流来美化自己的居住环境，在近些年，这对意大利的历史文化产生了不可估量的损失。各种水泥抹灰，粉刷不恰当的颜色和材质，装饰工业化帷幕，加上难看而无用的垫脚石，过度的重金属使用，都是对城镇历史文化保护的恒定威胁。

另一个严重的威胁是缺乏对切尔瓦拉镇特殊性与重要性的鉴定与认知，这非常地重要。比如知道一个屋檐的突出部分的尺寸，以便让这个装饰融入当地传统的建筑环境中，不让切尔瓦拉的房子与意大利北部特伦蒂诺的房子完全类似的情况出现；还有必要区分的是在正面和侧面的突出部分中，降低后者到零，因为这是当地的传统。不遵守这个简单的准则将会导致这个小镇整体形象的完全破坏。

历史记忆的丢失，阻碍了建筑使用者和工人对于众多元素的技术含义与价值形式的深入理解。古老的泥灰层迅速脱落，石头与石头的连接处分离，墙壁的古老泥层全部暴露在外，许多房子再也不是最初建造时的模样。如今很少有工人掌握传统的抹墙技术，不得不求助于工业的预混合材料，或者用别的适当的方法使外墙的石头平滑。从这种意义上来说，《修复指南》是强大的对照工具，技术解决方案。传统类型的价值及美观度应该引起房屋使用者、技术人员和工人们的注意，这是几百年来试验和选择的结果。

四、直接干预建筑结构以确保历史建筑的抗震安全性

与其他优秀的《修复指南》类似（比如有关城市堡垒的），《切尔瓦拉镇修复指南》也配备了适宜而简洁的有关抗震性修复的指导，这个指导基于分析程序和力求达到与干预标准一致的要求，并且符合现代抗震规范。这部分内容为修复策划者与需要采取措施来提高抗震性的工人提供一个有用的准备，以便更清楚地认识原有的建筑技术。

这个目标必须在保护原有基本结构的构想中完成，加强对艺术适宜规则的认识。从干预的角度看，欧洲标准是这样定义的：改善的措施表明，一个或更多的有关个体建筑结构元素的修复和实现最大程度安全性的目标的实现，基本上不能改变建筑的整体状态。相同的定义应当建立在干预类型选择采用的标准上，通常来源于对建筑结构的初步研究。关于地震地区特殊类型古迹修复干预工作的建议，由意大利全国文化遗产和地震风险委员会发布，他们认为涉及地震地区古迹安全的目标提前直接干预是最好不过的。

总之，切尔瓦拉镇有自然有传统，有历史有艺术，保护切尔瓦拉镇的一草一木、一砖一石，就是保护意大利的古老文明。

陈凌广
Chen Lingguang

浙江传媒学院设计艺术学院教授、副院长，校学术委员会委员，环境设计专业负责人，杭州师范大学硕士研究生导师，中国美术学院综合艺术系研究生毕业。

专著
2010年　《浙西祠堂》获浙江省高校优秀成果三等奖
2013年　《浙西霞山古镇民居文化及其时代价值研究》

课题
2014年　浙江省哲学社会科学规划课题《后家族时代浙江祠堂建筑文化艺术当代
　　　　价值研究》
2014年　浙江省非物质文化遗产项目《衢州祠堂营建技艺传承与保护研究》
2015年　浙江省社科联科普重点课题《浙西古民居精粹》

Graduated from Department of ComprehensiveArt, China Academy of Art, he is currently the vice-dean ofSchool of Art Design, Zhejiang University of Media and Communications (ZUMC), member of the ZUMC's academic committee, director of environmental design major, and graduate academic supervisor at Hangzhou Normal University.

Publications
2010　Ancestral Hall of Western Zhejiang,won the third prize of Excellent Achievements Among Universities in Zhejiang Province
2013　Study on the Culture and Its Contemporary Value of Xiashan Ancient Town Residents in Western Zhejiang Province

Research Projects
2014　On Contemporary Cultural and Artistic Value of Zhejiang Ancestral Hall Architecture in the Post-family Era, Zhejiang province philosophy and social science planning project.
2014　On Inheritance and Protection of Construction Techniques of Ancestral Halls in Quzhou, Zhejiang, Zhejiang province intangible cultural heritage project.
2015　Quintessence of Ancient Dwelling in Western Zhejiang, Zhejiang Federation of Humanities and Social Sciences Circles key popular science project

"留住一片乡愁"
——浙西霞山古镇民居文化及保护对策

陈凌广

[Abstract] In past thirty years in China, the economy grew at a rate of ten percent and the comprehensive national power was increasing everyday. At present, the country is during the process of transformation and upgrading, and the rise of economy needs to be boosted by culture more. Some people say that, Chinese contemporary culture is the integration of pre-modern culture and postmodern culture, and there is no modern culture and its value system construction. Where should Chinese contemporary culture go? As everyone knows, culture needs credibility, but Chinese contemporary culture lacks influence. Therefore, it is extremely urgent to solve the problem of lacking credibility of contemporary culture. China has very rich cultural heritages, especially the folk houses with the characteristics of ancient village culture carrying the elements of rich local humanistic feelings and the symbols of Chinese culture. Therefore, it is the only road to extract the elements of various cultures organically to transform into gene of contemporary culture for the construction of contemporary culture. The development of Chinese culture has entered a turning period. The author thinks that the construction of Chinese contemporary culture must return to transformation from "others' advices" to "using local materials". The choice of history has been clearly in front of us: the right way is only to take root in tradition, keep a foothold at present and look foreword to the future. However, the rules to construct contemporary culture shall not be made by us, but shall meet the standards of globalization and publicity. There have been very mature practices and general system internationally for the protection of various architectural cultural heritages. The author explained the profound mystery with the residential building culture protection in Xiashan Town located in western Zhejiang as an example.

中国在过去三十年，经济以百分之十的速度增长，综合国力与日俱增，如今，正处在转型升级过程中，经济的崛起更需要文化的助推，有人说：中国当下文化是前现代文化和后现代文化的集合，缺乏现代文化及其价值体系建构。中国当代文化何去何从？众所周知，文化需要公信力，中国当代文化影响力匮乏，因此，如何解决当代文化公信力不足的问题已迫在眉睫。中国有着非常丰富的文化遗产，尤其是以古村落文化为特征的民居文化承载着丰富的乡土人文情怀与中国文化符号要素，从中有机地提取各类文化要素转换成当代文化的基因是建构当代文化的必由之路。中国文化的发展已经进入转折期，笔者认为，中国当代文化的建构必须回到从"他山之石"到"就地取材"的转变。历史的选择已经明确地摆在我们面前：只有根植传统、立足当代、着眼未来，才是人间正道。然而，建构当代文化不是我们自己制定规则，而是规则要符合全球化、公共性标准。对于各类建筑文化遗产的保护国际上已经有着非常成熟的做法和通行的制度。笔者以浙西霞山古镇民居建筑文化保护作为案例，阐述其奥秘。

一、"迷宫霞山"文化资源概述

浙江衢州这个历来是边际交流最为频繁的区域，它有着浙、闽、赣、皖"四省通衢"独特的区位优势，霞山作为通往安徽徽州、江西婺源、浙江淳安的古驿道上的驿站古镇，一个浙西衢州山区的小小的聚落，历史上它曾隶属于徽州，几经转折，它是千百年来徽州文化与吴越文化碰撞的结晶。霞山古镇中的传统民居，依山傍水，黑瓦白墙，空灵俊秀，其饰以素雅的木雕、石雕和砖雕，充满着浓郁的历史文化底蕴。霞山古民居建筑文化的遗存，反映了古代霞山人对美与善、情感与理智、心理与伦理、艺术与典章等礼乐文化的独特审美取向。

著名乡土建筑保护与研究专家、清华大学陈志华教授在乡土记忆丛书的《总序》中写道："乡土建筑是乡土生活的舞台和物质环境，它也是乡土文化最普遍存在的、信息含量最大的组成部分。它的综合度最高，紧密联系着许多其他乡土文化要素或者甚至是它们重要的载体。不研究乡土建筑就不能完整地认识乡土文化。甚至可以说，乡土建筑研究是乡土文化系统研究的基础。乡土建筑当然也是中国传统建筑最朴实、最率真、最生活化、最富有人情味的一部分。它们不仅有很高的历史文化的认识价值，对建筑工作者来说，还可能有一些直接的借鉴价值。没有乡土建筑的中国建筑史也是残缺不全的。"（转引自罗德胤：《峡口古镇》上海三联书店，2009/04）建筑是人类文明的重要组成部分，古民居建筑是珍贵的历史文化遗产。它的可珍惜之处在于不可再生性，一旦毁坏则不可能复原。美国建筑学家简·雅各布斯在《美国大城市的生与死》中猛烈抨击了以物质功能决定论为导向的大规模拆旧建新，认为这严重破坏了以多样性为基础的城市社会文化生态。她在书中有一段精彩的论述："建筑遗产的再利用可以为传统中小企业提供场所，以增加城市的经济多样性与活力。"同理，这些古建筑都是一个城市或是地域性格的承载物，是文化的延续。

长期以来，霞山古镇形成了与之相适应的自然环境和历史环境，在漫长的封建体制下，它是"庙堂文化""士大夫文化""商业文化""市井文化""民俗文化"等的有效载体，而这些恰恰能够从某个侧面反映出浙西社会文化变迁的渊源和发展轨迹。"霞山古民居"及其文化遗产（包括古街道、古商埠、古钟楼、亭廊、祠堂、戏台、宅院、楹联、诗文、文献、古籍、高跷竹马、香草龙等），具有历史、文化、经济、建筑、民俗等多种价值，它是中国丰富多彩文化宝库中的瑰宝，也是浙西霞山人民创造性智慧的结晶。它不仅是明清以至近代特定历史时期社会政治、经济、文化发展的"活化石""活字典""活典籍"，而且通过这一"活化石"的保护利用和研究，还可以探析出包括霞山古镇自身在内和它对外所涉及的整个社会的曲折发展过程与发生的巨大变化。

浙西霞山古民居的历史文化源远流长，从选址、布局、结构和材料等方面，无不体现着因地制宜、因山就势、相地构屋和因材施工的营建思想，另外，至今仍然醇厚的民风、民俗依然在浙西产生重要影响。其整个聚落曾有民居、祠堂、商店、钟楼、凉亭、神庙等各类古建筑三百余幢，建筑外观造型优美，内部木构件雕刻精美，形象生动，是浙西古建筑雕刻中的精华。霞山古民居建筑技艺已经作为浙江省非物质文化遗产保护项目，在我省古建筑中独树一帜。

二、"伤痛霞山" 现状保护之殇

当历史在渐次递嬗时，文化总是在传承的基础之上出现或这或那的迁移与或多或少的变异。中国的传统文化准确来说是一种建立在农业文明基础之上的家族文化，家是小国，国是大家，维系家与国之间的媒介则是家族，任何家族的聚集与聚合都是以祖先崇拜作为信仰纽带，以祠堂作为祖先崇拜的物化形式与物质场所，从而形成鲜明的宗族意识与宗亲观念。霞山这个以郑氏和汪氏两大族群为特征的聚落，在近千年的传统文化繁衍中形成了两个具有典型特征的宗族文化，并以此为基础形成具有浓厚地缘性、政治性与宗法性的基层社会组织。坚守宗族文化与对宗族祠堂的膜拜，似乎成了每个人的集体无意识。而"后家族时代"的今天，人类社会已经从农业文明走向了工业文明，在一系列工业化、城镇化、经济化的浪潮中，加之诸如计划生育的实施、农业人口的城市化流动与迁移、城市化的扩张以及现代人个体意识的过度彰显与家族意识的淡薄等原因，家族故事似乎已成了如烟的往事，宗族观念也似乎成了一种可有可无的说辞，在许多年轻人的心目中，特别是当下的70后、80后、90后的心目中，伴随着家族祠堂与古民居的山村僻地无异于历史尘埃的废墟。霞山似乎进入了一个实至名归的"后家族时代"。

笔者曾于2000年开始就关注霞山的发展，起初古宅连片，街巷曲折，外人进入犹如进入迷宫，明代钟楼、郑氏宗祠、汪氏宗祠、郑松如故居、郑岩如故居、将军宅、山货店、古茶楼、古布店等建筑完好无损，这个因传统山货商贸、木材运输发展起来的村落集镇，在本世纪之初的那些年还是那样地被称作古意悠远的"古迷宫"。由于地处山区，耕地资源紧张，一批先富起来的村民在县国土资源局提供不出新的宅基地的前提下，冒然对即将倒塌的古民居偷偷实行地拆除重建，当地村干部也因为怕出安全责任事故未及时阻止，尤其是在本世纪初的10年间，一批无知的村民看不到历史建筑所蕴含的丰富的资源价值，在经过近十年无序的建设，古镇的四周和村落的中心的现代建筑犹如炮楼般矗立着，一块块撕裂着古镇的肢体……原先近两百多幢明清至民国的古建筑只剩不到百余间。小洋房挤垮了黛瓦粉墙，暴露了古建保护的无奈，撕裂了一段悠远的感情，击碎了近一代人的乡愁！

三、"文化霞山"——古镇当代价值

继承与创新是当下最为宝贵的精神与诉求，一个古村落的保护固然是对古人生活方式的原生态的传承，更是民族建筑装饰形态之物本的存在和发扬。就传统资源而言，若安转换为一种当代的建筑实物存在，它既需要寻找到传统建筑形式上的当代因素与材料，技术上的当代运用及拓展，又要使传统建筑的形态和功能找到与当代生活方式的紧密契合，这才能真正让传统文化资源落地生根，发挥出当下时代的强音。在建筑设计领域，"文化是根"，在建筑装饰设计时将当地的人文、风俗、生活习惯、地域特色等软文化要求与建筑装饰，设计进行有机融合，尤其是与建筑文化相互关联的民俗习惯、道德规范、理想情操、宇宙观等地域性的意识形态相契合，在不断的联系交往中沉淀下来的本地区的地域性文化，往往反映在地域性的建筑文化上。在古民居建筑装饰中，不同的符号、材料、色彩、形式、空间组织和景观方面都可能包含某种文化意义，当意义、空间与活动系统相互一致时，彼此之间就加强；当建筑变得与社群文化和生活方式一致时，就有归属感。灵活地运用当地的地方性材料，使得所设计出来的建筑风格一方面能符合本质、节俭、节约的原则，就地取材，另一方面也能够充分体现出地方特色来，使得地域文化在建筑装饰设计中得以充分释放与发扬。

另外，伴随着几百年来世世代代霞山人民流传下来独具特色的"高跷竹马舞""板龙灯""花草堂"、"香草龙""跳魁星"等民间艺术活动，如果进一步对其进行发掘、整理研究，将有助于浙西传统文化遗产和非物质文化遗产的保护和利用，有助于文化的交流与传播，有助于文化的再创造，是体现传统文化遗产的文化价值的重要途径。

就霞山古镇的时代价值而言，它处在一个"历史的处境"中，这个历史是难以逾越的时间节点，受到特定的语境和文化影响，有一种被现代城市文明所背离的所谓"淘汰"的物质环境与诉求。可是正是这样一个视域的表象，却孕育着太多人们对过去农耕时代先民生活方式的记忆，这种记忆稍不加以珍视，残酷的现实就会像癌细胞一样快速地蚕食脆弱的"古民居"，一切的经典将无从说起，一切的梦幻将不再重现。

四、"未来霞山"——保护之机

安徽歙县的宏村、西递，浙江的兰溪诸葛村、江山廿八都、桐乡乌镇的整村保护已经为霞山的保护开启

了一个成功典范。我们处在一个民众审美提升、政府保护意识觉醒的时代语境下，尤其是国内如"猪栏酒吧民宿""栖迟艺术酒店""山水谈民宿"等一批精品村落民宿的成功，让我们完全有理由相信，保留历史和地域的符号、化腐朽为神奇，让没有生命、濒临倒塌的历史建筑，焕发新的生命，让有记忆、有沉淀的古村民宅成为心灵放归地，成为人们乡村"家"的记忆，去共同承载那一段乡愁。

古宅不仅承载着一段悠远的历史和情怀，更有效地将本民族传统精神与特色文化的有效内容继承下来。对待古建筑首先要充分地尊重原有场地、原有建筑，保留体现原建筑的历史价值，在此基础上创新、腾笼换鸟；其次做到空间调整包含外部空间的优化与重组、内部空间的充实与置换、新业态的植入、新旧功能的转化、空间的转化；其三处理好"新与旧"的关系，修新如旧、修旧如新。在运用古民居文化资源的创新上，如何将丰富的文化内涵与形式语言转化成当代文化，笔者认为更应该站在批判与吸收的立场上做到以下四点：1.传统形式的当代应用；2.技术运用与手段现代；3.生存经验与当下关联；4.传统与当代文化意识的共存。

参考文献

[1]《建筑的生与死——历史性建筑再利用研究》，陆地 著，东南大学出版社204.01。
[2]《历史城市保护学导论——文化遗产和历史环境保护的一种整体性方法》（第二版），张松 著，同济大学出版社 2008.03。
[3]《浙西霞山古镇民居文化及其时代价值研究》，陈凌广 著，中国电影出版社2013.11。

（本文是2014年度浙江省哲学社会科学规划项目"后家族时代"浙江祠堂建筑文化艺术当代价值研究段性成果）

劳伦特·莱斯考普
Laurent Lescop

法国南特高等国立建筑学院建筑教授，1966年出生，1999年获得建筑科学博士学位，供职于南特高等国立建筑学院。计算机图形专家，致力于推动历史遗产保护的发展，将历史遗迹视为一处景观和尚未完成的艺术品，工作涵盖了从考古研究到研究科学数据的各个方面。

2000至2001年，根据自己的建筑师工作经验，开发了用于建筑设计的全新电脑工具，获得了创新奖。在教学和科研中，一直致力于使用沉浸式互动体验设备和实时软件，推动"叙事设计"的发展。现在受到质疑的布列塔尼地区的地标建筑都应用了这些原则。同时他撰文讨论柏林墙和南特城市中心的转变。

ENSA-Nantes, France

Laurent Lescop, born in 1966, an architect with a PhD in architectural science (1999), is a computer graphics specialist, teacher and research professor at the Ecole Nationale Supérieure d'Architecture de Nantes, France. His work develops the concept of heritage, considered as a scenic place and works in the process, going all the way through from archaeological survey to communicating the scientific data..

15 years ago, he developed new computer tools for architectural design (Innovation Award), that he incorporated in his own practice as an architect (2000-2010). He tries to develop, both in his teaching and in his professional practice or in his personal research, the idea of "narrative design" based principally on immersive devices and real-time softwares.

Those principles are now applied to iconic sites in Bretany where heritage is questionned. He also wrote on contemporary situations like the Berlin Wall or the mutation of the city center in Nantes.

保存树根以孕育森林
——南特案例研究

劳伦特·莱斯考普

概要

我们从来没有在欧洲试验过中国现在在进行的大规模城镇化。但是，急剧的经济转型迫使决策者们开始从发展的角度重新思考城市布局。当前出现的问题有：就业、住房、交通、医疗、教育、停车和运动场所。人们关注的方面包括：新科技、健康、生态和港口。

同时，城市本身也在寻求自己的品牌和身份，这是一种可以用来吸引投资的商业优势，城市居民也可以从中获得一种认同感。如果城市品牌主要是源自该地拥有一家著名企业，比如斯图加特的奔驰，索绍的标致以及马拉内罗的法拉利，那么我们就需要从文化和遗产入手。

对于像巴黎、罗马和雅典这样历史底蕴深厚的城市来说，这从来都不是个问题。但是没有那么丰富文化遗产的城市就需要研究这个问题了。

而且，文化遗产的定义也发生了改变。保存某地某处建筑遗迹的历史风貌已经不是唯一保护建筑和文化遗产的方法。我们所参与的当地历史和生活习俗都是文化遗产。每个人都应该意识到自己就是历史进程中的一部分。

因此，文化遗产不仅是指建筑，更多是指所谓的无形遗产：民众、工匠、艺术家、各行业的工人、农民、音乐家和教师。他们的知识必须得以传承，并且根据新时代的趋势和项目有所改变，以孕育新的发展。

南特在1986年关闭了它的最后一个船坞，之后就陷入了困境。象征其工业文明的起重机接二连三地停止了工作。失业率攀升。拆除工厂并翻新城市或者努力保存

在2000年后的头几年，南特老工业区的复兴工作给了人们一种新的思路。竞赛获奖者Alexander Chemetoff提出了一个融合历史和未来的城区规划，而不是将这个区域统一规划为住宅区。工业文化遗产因此通过两种方式保存。建筑得以翻新，其原有的结构不发生变化。而与之相关的知识，也就是无形遗产，被重新注入到文化产业中。

这样一来，南特在形象和品牌上完成了从工业城市到文化城市的转变，发展了以高科技为主的新产业。城市面貌焕然一新，这就是我们法语中说的" art de vivre"（art of living）。

在法国，我们的主要目标之一——是在每个城市恢复乡村生活，这种情况很常见。"第二次世界大战"期间，美国士兵有一本袖珍书，以便于更好地了解他们将会遇见的人们。对于法国人来说，他们的生活中只有一个目标：在花园里种植作物（Collectif，2004年）。我认为这就是为什么法国人在每种情况下尽力尝试重新创造在乡村、在花园里种植自己的食物的理念。甚至在大城市也是这样，人们常常会听到"巴黎是一个村庄"这句话。正如电影《艾蜜莉》（Jeunet，2001）描绘的那样，每个社区都保留了前村的记忆。艾蜜莉在巴黎的生活方式真的就像是在乡村里生活一样。她在街角购物，每个人都相互认识，每当艾蜜莉离开家，她就像去旅行一样。

要成为总统，植根于土地同样重要。弗朗索瓦·密特朗竞选总统时，他使用的就是这个理念。在1981年竞选活动期间，他的海报显示的是他站在一个小村庄中与村民们在一起的背景。结果，他竞选成功了。相反，法国前总统瓦勒里·季斯卡·德斯坦为了表达现代性，在海报中使用了工业背景，竞选失败。讽刺的是，1969年，密特朗也在海报中使用了工业背景，那是他第一次参加竞选，结果与德斯一样：竞选失败。2002年，尼古拉·萨科齐在海报中也采用了乡村形象，这一年就是他竞选成功的一年。五年后，萨科齐因采用了海边风景的海报而输给了采用乡村风貌作为背景的弗朗索瓦·奥朗德。

法国电视台的许多电视节目都是以国家最好村庄的发现或选举为基础的。村庄是指小规模的社区，表

明人们在一个社区，彼此互相认识，相互帮助。一个村庄可以看作是一种特定生活模式（亚历山大，1978年）。村庄反对"城市群"，个人反对集体，弱者反对强者，供应商反对消费者。这样看待事物的方式也是因为村庄变得荒芜、被遗弃，而强调了遗产问题。但遗产的定义已经转移。遗产不仅是建筑物，越来越多的是现在所谓的非物质文化遗产（联合国教科文组织《文化》非物质文化遗产，s.d.），是指工匠、艺术家和许多工人、农民、音乐家和教师。他们的知识应该被传达，也应该适应新的需求、新项目，成为新发展的种子。"你没开玩笑吧？"，我们在谈论自己的城市和象征其发展和财富的大象的时候经常会听到这样的评论。我们确实没有开玩笑。2004年南特被命名为欧洲最宜居城市：毗邻大西洋，卢瓦河流经于此，周围遍布酒庄，与巴黎之间的交通便利。南特是一个富庶且充满活力的城市，法国西部的其他城市也是这样。

这当然是一次伟大的复兴。之前很久的一段时间，人们很难定位南特。这虽然是个港口城市，但是远离海洋。人人都害怕卢瓦河。这座城市的两片区域由桥梁连接，但是只是向北部发展。这是一片巨大的城区，但是中心没有得到开发。

从幻想到现实

我们可以从一个城市的明信片、旅游手册或者电影中察觉这个城市的形象和气氛。摄影师通过电影来描绘他眼中的城市，这通常能反映市井生活，不管是好是坏。比如战后的罗马是一个令人绝望的地方，Roberto Rossellini 1945年的电影《罗马，不设防的城市》以及Vittorio De Sica1948年的电影《偷自行车的人》呈现了一个充满贫穷和暴力的战后罗马。William Wyler 1953年拍摄的电影《罗马假日》则充满了轻松的气氛，主人公在明媚日光中游览著名景点。Woody Allen2012年的《爱在罗马》将罗马描绘成一座永恒之城，一张旅游明信片和一个露天博物馆。很多电影也是在法国取景的。南特虽然不如巴黎或者尼斯或者马赛在电影界这么有名，这个城市还是在二战以后成为了大约20部电影的取景地。电影《罗拉》或者Jacques Demy拍摄的《城里的房间》为访客们营造了一个虚幻现实。1961年拍摄的《罗拉》展现了一个重工业城市南特，造船厂规模巨大且起重机遍地。船坞中到处都是强壮的劳力，街头巷尾都可以看到妓女。港口显然不适合遛狗。

图1 Lola, J.Demy 1961

1982年，南特出生的Jacques Demy 拍摄了《城市中的房间》。场景设置于1955年，描绘了劳工们对未来的迷茫。但是在1982年，部分未来已经成为了现实。毁于1958年的标志性的输送桥通过遮景绘画手法得以重现。南特的特色就是起重机，船坞和工人。

　　1986年，南特关闭了最后一个船坞，这个曾经富庶的城市陷入了困境，纵然它的财富是来自曾经的奴隶贸易。这座城市迷失了自己的身份，当地足球队的胜利能够偶尔提升下城市的知名度。5年以后，法国导演Jean-Loup Hubert和Catherine Deneuve拍摄了《白色女王》。该片取景于卢瓦河南岸的一个叫作Trentemoult的小镇。该片充满了对南特鼎盛时期的怀念。河流另外一边远处起重机的轮廓透露出浓厚的怀旧情怀。10年以后，Pascal Thomas于2001年在南特拍摄了《疯狂星期三》。片中的城市活泼时尚，主人公搭乘公共交通出行，快速地穿梭各地之间。但是起重机仍然是该片的主题，它们就像是古代怪兽的骨架，将长长的獠牙托向天空。片中的起重机比现实中的更大，从任何视角都能看到。但同时发生了什么呢？

　　1989年,Jean-Marc Ayrault在当选市长之初就举办了一场鼓励重新修缮市中心的比赛，但当时市内的居民还是很怀旧。胜出的队伍带来了很多新想法，包括减少私人用车，调暗街道灯光并将大片区域改建为草地和公园。

　　南特在历史上曾经被分为好几个岛，东西向被卢瓦河分开，再由垂直汇流到卢瓦河的埃德尔河从北分开。渐渐地，岛屿融合在一起，河岸线也逐渐平直。到了20世纪，支流和埃德尔河都被填平了，留出大片空地，变成了街巷，道路和停车场。汽车逐渐退出了城市中心。由此，空间分割导致了社会分区。

　　但是南特在20世纪90年代发生了剧变。这种剧变不是经济层面的，而是发生在艺术界。 1990年至1995年之间，名叫"Les Allumes"的艺术家团队每年都会点燃城市的夜空。他们在1990年10月15日到20日之间发起了第一次这样的活动。 Jean Blaise首先创办了这个自由户外戏剧活动，该活动会邀请"一个国外城市"来做客，第一年他们请来的是巴塞罗那。活动举办期间，南特各处都会组织故事会、演唱会和戏剧表演，城市顿时成为欢乐的海洋。接下来的几年主办方还邀请了圣彼得堡、布宜诺斯艾利斯、那不勒斯和开罗。活动还邀请了著名的艺术家在城市各处举办展览。印象古典音乐(La FolleJournée)节也是参考这个模式举办的，它为大量的观众带来了顶级古典音乐会。

　　推广街头剧场的目的是吸引并引导不同层级的观众。1978年创建于法国南部的戏剧公司Royal de Luxe 于1990年在南特举办了一场声势浩大的"法国的历史"游行。大型的机械道具和盛装的演员们乘坐着的彩车巡游城市，到处欢声笑语，人们大肆庆祝。游行取得了巨大的成功，该公司也顺理成章地参与了城市的翻新工作，居民们也很高兴地加入其中。

　　Jean Blaise 在2007年回顾这个时期的时候说到： "只有很少的城市像南特一样用文化撬动了经济的

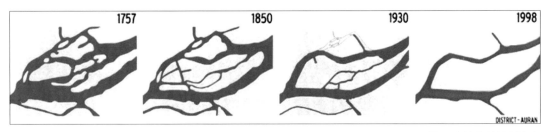

图2 南特岛的地形演变

发展，当时很有挑战性，但是今天取得了成功。我们寄希望于用国外的和风格独特的活动来象征南特市如同"Les Allumes"般的转变。全国的媒体趋之若鹜。城市的形象在一年之内完全改变。Royal de Luxe 之类的著名艺术家和舞蹈艺术家Claude Brumachon纷至沓来。"和南特一样"就是指在展示高雅艺术的同时也取悦普通大众。比如，我们在南特提供很多机会去重新发掘城市的工业遗产，但同时也去开发像"ile de Nantes"这样的新社区。文化可以促进经济和社会发展，文化也可以用来探索未知领域。在"Lieu unique"，我们每年至少接待49万访客。这里并没有大型的节日，只有一系列小型活动。我们也接纳艺术家。我们希望把双年展"港湾2007"发展为一个地区性的项目，覆盖从南特到圣纳泽尔的区域。这个区域已经是一个经济共同体。工业对该地区的经济发展做出了重要贡献，曾经密布工业设施的港湾地区现在将被改造为步行区域。

图3 埃德尔河填河之前

图4 改建前的车流

城区规划和景观规划

　　1992年12月，建筑师Dominique Perrault在南特开展了一次探索性的研究，并为城市改造制定了一些主要原则。在名为《方法》的文章中，他认为为一大片区域的发展制定一个总体规划是不现实的。改造必须针对"不同且并存"的问题，不分先后。他建议推广新的安装、连接和参与方式。在2000年之前，南特的城市发展并没有层次的覆盖。每个发展时期都会开辟一片新区域，而后该区域会被发展或者废弃。这就导致南特的城区高度区块化，边界非常明显，规模差异巨大。最终，位于南特岛中央的工业区块变成了一块荒芜危险之地，它就好像是城市中心的一块死区。

图5 今天的街景

　　我们来了解下具体的情况。岛的东部是工业发展留下的大块空地。岛的中部居住着低收入人群。岛西部有些新建筑，但是绝望的居民希望尽快离开此地。南特岛被一条高速公路分割为三部分，这条高速公路使岛南的居民可以直接驾车到达中心地区。Perrault未能明确回答如何处理这大片工业遗迹以及落后的城镇化问题。电影将南特描绘成一个起重机之城。而对于当地居民来说，自己的城市并没有清晰的形象。当地开展了

图6 规划图 (Chemetoff, 2010)

一项关于南特形象的调研，出人意料的是，大多数居民并不知道南特是一个完整的岛。很多人认为南特还是由小岛连接起来的，这可以从当地的地名中找到原因。有人说他不能绕着岛遛狗，所以这肯定不是真的。岛东部被称为Beaulieu，这里是一个重要的住宅区。对于其中四分之三的居民来说，这是一个中转站，70%的人会在居住了6个月之内搬走。如同建筑师所推崇的那样，办公区混合在住宅区中，这样可以让人感觉这不是一个私密区域。

如我们所见，在2000年初期，南特面临严重的城区规划困境。我们可以从电影中得知主要的文化遗迹位于工业区域——港口，而那时港口已经关闭。关于起重机、船舶和生产车间的知识、形象以及景观都已经在20世纪80年代中期的危机中消逝。南特在1999年发起了另一项重新定位港口区域和现被称为"Ile de Nantes"区块的竞赛。尽管经济学家们极力推荐参考毕尔巴鄂的模式，最终获奖的方案选择了一种"低调"的方法缓慢改进城区，而不是建造一个巨大的城市地标建筑。加拿大裔美国建筑师Frank Gehry设计的毕尔巴鄂的古根海姆博物馆于1997年10月竣工。从此以后，似乎出现了一个工业保护、文化发展和经济改善的配方：一座巨大的宏伟建筑，一个环境优美的投资区域和来自全球的名家的艺术作品。南特并不赞成这样的做法，这其中有哲学的原因，也有出于经济的考虑：一个高失业率的城市无法花巨资建造一座并不一定能够确保经济可持续发展的建筑。由Alexander Chemetoff带领的获奖团队在其初稿中提出他们将像考古学家那样寻找隐藏的近代历史轨迹。他的设计保留了古代工业建筑和当地民间知识。

有三个建筑师团队最终入围：Labfac，Brubo Fortier和Chemettof。Labfac设计了一个绿色的正弦形，Bruno Fortier设计了一个绿色的走廊。Fortier说："我们必须给予南特岛一个大都市级别的有力并且时尚的形象。"他认为建筑就是一个巨大的博物馆，就比如毕尔巴鄂的建筑一样，成为一片绿地中的靓丽景观。

Alexander Chemetoff则采取了一种完全不同的方法。他将南特岛划分为公共区域和私人区域，并在"河岛"和"海岛"之间共享。Chemetoff提交的规划原则将使得公共和私人投资者了解到项目概况。南特的主要问题在于，该区域中城区空间和建筑的历史价值并不明显。

区域内没有特定建筑会被保留。建筑师和当地协会就建筑的拆除和修缮问题讨论了很久。最终，大家都认为这个区域是一块保留了当地知识、历史和生活方式的无形历史遗产。难点在于将其转化为一种城区设计。Alexander Chemettof首先从效果图的视觉效果入手。他把效果图画得和游客中心的旅游地图一样，项目看起来就像已经完成了一样。然后，他开始着手分区。工业区到处都是大型建筑，没有市区规划。他重新规划了街道、广场和游憩场所。对他来说，不管这些区域最终用途是什么，他只是把它们划分为公共/私人区域。Chemettof接着开始像一个考古学家那样工作，发掘并保存所有的有形历史遗迹。当然他也保留了旧港口里地标般的起重机。有些工业建筑得以翻新，人们保留了主体结构，但是拆掉了其他部分。新建部分易于拆除，没有什么东西会一直保留。

最早的想法是开办一个生物科技创业园。这最终没有成形，但是即便没有落实，对于城市的复兴来说这也是个很不错的建议。但是问题在于这样做会失去城市的无形遗产并脱离了当地的工业传统。与此同时，大型机械道具游行取得巨大成功，这使得Chemettof想到了另外一种延续传统的方法。当地的港口工人仍然可以从事他们以前的工作，但是这次目标完全不一样。他们不是造船，而是建造巨大的大象。

巨大的木象和巨大的小女孩

2005年，巨型木偶完工：小女孩和Sultan prince 以及他的大象。场面非常壮观。我们很难想象工匠们是如何将他们的奇思异想融合到一个巨型木偶中去的，但是目睹一个和城区最高建筑一样高的大象在惊呼的人群中穿过街道，这实在是让人难以忘怀。木偶取得了巨大的成功，市民们非常喜欢这些木偶和它们的创作者。城市重新焕发了活力。

在毕尔巴鄂建成了其地标性博物馆并翻新市貌之后，人们认为文化可以为寻求复兴的城市带来价值。在新世纪的头几年，因特网和电脑的普及为南特带来了繁荣和经济增长。与此同时，街头剧场，尤其是木象，已经开始吸引越来越多的游客到访南特，有些甚至在此定居，市民们也因此深爱自己的城市。这也带动了外来人才的流入，促进了经济发展。

在2000年中期，从Royal de Luxe分离出来的戏剧公司La Machine 新建了一个后来成为南特emblem的大象。这个大象并不是一件艺术品，而是一个可以移动的建筑。这完全颠覆了人们对市区的看法。在某种程度上来说，毕尔巴鄂效应并不来自于一个地标性的博物馆，而是来于一个巨型木象。这个大象吸引着游客和市民来到南特那片之前的"禁入"之地。在大象之后，艺术家兼设计师Francois Delaroziere也创作了很多机械动物，有些是为南特建造，有些用于游行，还有些是为其他城市建造的。

从外人看来，完成这样一个项目并且将文化和工业遗产融合在一起是非常令人吃惊的。遗产可以是建筑，南特保留了主要的地标建筑以保留历史。遗产也可以是一种无形的东西，就像传统和食物这些。在南特，遗产最终不是保留特定物品，而是用新方法在原来的历史上创作新的故事。成功也造成了困惑和疑虑。有人提出"南特方法"，还有人开始仔细研究其中的过程。但这之前没有任何相关的研究和定论。城市规划和建筑层面都是从草稿开始一步步完成的。 Chemetoff说他是在一个模糊的概念上开始逐渐聚焦。形象越鲜明，所呈现的细节就越多，细微差别就更加易于识别。"鲜活的逻辑"是一个非常强大的概念。Chemettof说："关于城市的所有工作都要求你能够接受一定的不确定性，这就和种树一样。你选好了地点和树种，但是你不知道它到底会如何生长。城市就是一棵树，是一个鲜活的个体 。"

南特在十年内根本改变了城市形象。为了达到这个目标，它需要创造很多新的画面。在21世纪初期，一个城市的新范本就是能够产生很多可以在网上传播的视觉作品。 Jean-MarcAyrault当选南特市长的时候，南特正在改变自己的形象，以前的南特如电影所描绘的一样：遍布静止不动的起重机，如这个城市般死寂。但是后来出现了街头剧场，人们开始走出家门，带来新鲜的想法和视野。 Jean-MarcAyrault由此发起此项开发荒废区域的比赛，他明白了可以将市区规划和街头剧场联系起来，这是一个新思路，是可行的。

图7 第一头大象

南特的居民们开始逐渐熟悉大胆的外形设计和现代建筑，他们甚至在周日和节假期蜂拥围观木质大象。我们当然有些开始遗忘起重机了。它们是承载港口历史的遗产，或者只是映衬机械动物的蒸汽朋克风格背景图。南特岛开始自我发展，它不再是城市的累赘，恰恰相反，它已经成为了城市的发动机和灵感。十年后人们很难想象这个地方会变成其他什么样子。人们开始逐渐遗忘木象游行和城市翻新之前的景观。南特由一个遍布起重机的工业城市变为以木象为标志的文化之城。

图8 新的大象

别的城市也可以这么改造么？比如在法国的其他地方，或者欧洲甚至是中国的其他地方？南特最大的竞争城市波尔多也使用了相同的方法，并发展出了我们现在所说的叙事性都市设计。南特附近的拉罗什邀请Chemettof在其一个非常有争议性区域使用相同的改造方法。Chemettof请木象的创造者François Delarozière编著一个能够广受欢迎的故事。François Delarozière设计了一些就像木象一样的机械动物，民众都可以进行操作。机械动物在全球都受到了欢迎。2008年利物浦的巨型蜘蛛，

图9 新的巨龙

2010年的空中温室以及去年在中国北京的龙马精神，都是其中的著名作品。就像在南特一样，在区域发展、可持续性和民众福祉等问题上，我们可以想象这些非凡的生物可以成为连接过去和现在的纽带。

事实上，最主要的问题是"什么是遗产？" 欧洲目前的主要做法就是冻结物体的状态，把它变成一个没有生命的展品。古代城市和工业遗迹都变成了露天博物馆。这种做法迅速普及到全球其他地方。但问题在于，其他地方会把其中的居民也一并迁出。另外一种想法也提出要保护无形遗产。现在联合国教科文组织/

国际古迹遗址理事会正在寻找新的途径向其中注入生命，艺术，创新和创造精神，以使这些古迹能够不仅呈现其原有面貌及其所代表的事物，更能使参观者有所收获。

　　艺术已经成为城市发展的一方面。就像蒙特利尔一样，现在南特的心脏就是"Quartier de la Création"，意思就是"创造区域"。这里有教授艺术、建筑、设计、时尚、舞蹈和电影的学校，不断地吸引着各种工程师和科学家前来。"区域"也可以理解为"村落"。南特岛的西部将成为一个创意村落。这是一种别有趣味的开启新时期的辞旧迎新。

Shaping the image of a city Nantes a case study

Laurent Lescop

Summary

Nowhere in Europe, have we experimented such growing urban scales that we see in China. However, violent economic shifting pushed decision makers to rethink cities with a new aim of development. Several strong topics are mobilized such as: employment, housing, transportation, hospitals, schools, parks and sport fields, with the focusing on themes like: new technologies, health, ecology, harbors.

Cities are also looking for a branding, an identity, that could be used as a commercial advantage to attract business, but also help inhabitants to share something in common. If city-branding has been generally based on the presence of a huge company such as Mercedes in Stuttgart, Peugeot in Sochaux, Ferrari in Maranello, this concept moves to culture and heritage.

For cities with a strong historical background, it has never been a matter of discussion: Paris, Rome, Athens is clearly identified, but for cities with less famous or abundant heritage, this question may be an issue.

Moreover, the definition of heritage has moved. Places, monuments, buildings to protect are not the only way to conceive monumental and cultural heritage. We now work on everyday life consideration, local history, craftsmanship, and habits. Everyone should have the feeling to be part of the process of becoming part of it.

Consequently, heritage is not only buildings, but more and more what is called Intangible Heritage: that means people, craftsmen, artists, and workers of many kinds, peasants, musicians, and teachers. Their knowledge should be transmitted, but also should adapt to new needs, to new projects, be seeds of new developments.

In Nantes, the last shipyard closed in 1986 leaving the city in a desperate situation. The cranes, symbolizing the industrial activity, one by one stopped. Unemployment stroked. The question was between turning the page, tearing down the workshops and reinventing a new story or trying to preserve would appear to most of the population, a kind of modern bulky legacy.

In the early 2000's, the revitalization of Nantes' former industrial area, led to develop a new way thinking. Instead of designing an urban map with major spots and rows of housing, Alexander Chemetoff, winner of the competition, thought better to draw an urban landscape where the past could mix with the future. Industrial heritage has been then preserved in two different ways. Construction halls have been reshaped preserving the original structure, everything should be reversed. The intangible heritage, meaning worker's knowledge has been reinvested in the cultural industry.

This way, the image of the city, its brand, moved from industrial to cultural, attracting a new kind of business, mainly high-tech, students, in a new, what we call in French: "art de vivre" (Art of living). It also moves from metropole to village by the subdivision of huge entities.

It is common to say that in France, one of our main goal is to retrieve a kind of village life in every urban situation. During the 2nd World War, American soldiers had a pocket book to better understand the populations they would meet. For French it was written that they would have only one goal in their life: being in their garden (Collectif, 2004). I believe that's the reason why, French try so hard to recreate, in every situations, this idea of being in a village, being able to garden, to grow their own food, to be deeply rooted in the ground. It's true even in big cities: "Paris is a village", is a common sentence. Each neighborhood hold the memory of the former village as it is beautifully pictured in the film "Amelie" (Jeunet, 2001). The way she lives in Paris, is really like being in a village. She goes shopping round the corner, everybody knows each other, and every time Amélie leaves her home, she's like having a trip.

To become president, it's also important to appear rooted to the ground. François Mitterrand used this idea when he ran for presidency. His poster, during his campaign in 1981, showed him standing in the countryside with a small village in the background. And it worked, he was elected. On the opposite, to express his modernity, former president Valery Giscard d'Estaing used an industrial background for his poster, and he failed. The irony is that Mitterrand also used an industrial background on his poster in 1969, the first time he competed and he got the same result: he lost. In 2002, Nicolas Sarkozy also took a countryside image for his poster, that's the year he has been elected. Five years later, he took a seaside landscape and lost against François Hollande who had a countryside landscape for his background!!

Many TVshows on French television are based on the discovery or the election of the nicest village of the country. Village means of course small scale. It suggests that people are in a community, they know each other, and they can help each other. A village can be seen as a pattern (Alexander, 1978), including a certain way of life. The village is against "metropolisation", it's the individual against the collective, the weak against the strong, and the provider against the consumer. This way of seeing things is also because villages are deserted, abandoned, stressing issues on heritage. But the definition of heritage has moved. Heritage is not only buildings, but more and more what is now called intangible heritage (UNESCO » Culture » Intangible Heritage, s.d.): that means people, craftsmen, artists, and workers of many kinds, peasants, musicians, and teachers. Their knowledge should be transmitted, but also should adapt to new needs, to new projects, be the seeds of new developments.

Fiction to understand reality

"Are you serious? " It is a very frequent comment we hear when we talk about our city and its Elephant, which is seen as a symbol of development and welfares. And yes, we are serious. In 2004, Nantes was named "the most livable city in Europe": lying close to the Atlantic Ocean, beside the Loire River, surrounded by vineyards and well connected to Paris, Nantes is a prosperous and forward moving city, just ike any other city in western France.

It has certainly seen a great revival. For a long time, the identity of Nantes was difficult to picture. It's a harbor, but far from the sea, it's a river that everybody fears of, it's a bridge city, connecting two regions, but only develops towards the north, it was a huge urban area, with a void in its center.

To understand the image of a city, the ambiance, one can look at postcards, tourist guides or cinema. Cinema is a good way to see how a city is perceived from a cinematographer point of vue, it often reflects what is really in the air, positive or not. Let's take Rome, for instance, after the war, it was a desperate city and films like "Roma città aperta" (Roberto Rossellini 1945) or "Ladri di biciclette" (Vittorio De Sica, 1948) show a postwar Rome of poverty and violence. In "Roman Holiday" (William Wyler 1953), the atmosphere is lighter, the sun shines, the characters ride through famous places. Woody Allen in "To Rome with Love" (2012) shows the eternal city, it's a tourist postcard, an open-air museum. A lot of films are also shot in France. If Nantes is not as famous as Paris or Nice or Marseille for film making, about 20 films were shot since the end of the Second World War. Famous movies like "Lola" or "une chambre en ville" by Jacques Demy, created a fictional reality that visitors wanted to find. "Lola" filmed in 1961 shows Nantes as a city of cranes, with a strong shipyard and a dense industrial activity. Docks are full of strong guys, ready for a fight, prostitutes and lost girls. The harbor is certainly not the place to walk the dog.

In 1982, Jacques Demy, who, by the way, was born in Nantes, shot "Une Chambre en Ville". The action is set in 1955 and it shows laborers wondering about their future. But in 1982, part of this future is already known, and the iconic Transporter Bridge, destroyed in 1958, is recreated with a glass shot (Aumont & Daguin, 1998), a matte painting. Nantes is characterized by its cranes, the shipyard, and the workers.

In 1986, the last shipyard closed, leaving the once affluent city, even if its wealth did come from the slave trade, in a desperate state (Barrau & Wester, 1998). The city had lost its identity, which was only boosted up once in a while with its football team. Five years later, French director Jean-Loup Hubert, filmed "La Reine Blanche" with Catherine Deneuve. The action takes place on the south bank of the river Loire, in a small village called Trentemoult. In this film, everybody can feel the profound nostalgia of a lost era when Nantes was strong, powerful. This nostalgia is shown with the silhouette of distant, unreachable cranes, on the other side of the stream. Ten years after, in 2001, Pascal Thomas shows Nantes in "Mercredi, folle journée!". The city is active, modern, characters are using public transportation. They are moving fast from place to place. But still, cranes remain as a leitmotiv, like the skeletons of ancient monsters, rising their tusks up to the sky. Cranes appear in this film bigger than in reality, they can be seen from any place.

What happened meanwhile?

1989 saw Jean-Marc Ayrault, the newly elected mayor, set up a competition to encourage the rebuild of the city center, but the city was still feeling very nostalgic. The winning team succeeded in bringing new ideas, including the reduction in the number of cars, dimming of street lights and converting large areas to grass and

Fig. 1 Lola, J.Demy 1961

parkland.

In its past history, Nantes was divided into several islands, scattered east/west by the river Loire and split by the river Erdre coming from north and joining the Loire in perpendicular. Little by little, the islands were merged, drawing a more linear riverbank. In the 20th century, the last branches were filled as well as the river Erdre, leaving huge empty spaces, taken by cars as roads, streets or car-parks. Little by little, cars fled the city center. As a result, spatial divisions caused social partitions.

But in the 1990's saw a great upturn in Nantes. It didn't come from the economy, but from the art. Between 1990 and 1995 the artist group "Les Allumés" annually lit up the city's nights (Saranga, 1993), with the first event running from 15th to 20th October 1990. Dreamt up by Jean Blaise, this free, open air, theatrical festival was designed to "invite" a foreign city to Nantes, that first year being Barcelona. Story-telling, concerts and plays, were performed all over the city turning Nantes into an exhilarating place. The following years saw Saint Petersburg, Bueno Aires, Napoli and Cairo on the guest list. Great artists also joined the festival, exhibiting all over the city in venues ranging from large hall to small bars, in the open air and in private houses. The classical music festival "La Folle Journée" was also organized in the same manner, and brought top quality concerts to large audiences.

The idea of promoting street theater as a way to get people of different ages, interests and education levels together then followed. The theatrical company "Royal de Luxe" established in 1978 in the south of France took, in 1990, Nantes by storm with a thunderous parade of the "History of France". Dozens of actors in costumes paraded through the city on large carnival floats, with huge mechanical sets, blasting objects, creating smoke in the streets, shouting and cheering. The success of the parade was considerable and there was no question that the company was now part of the renewal of the city, and the residents gladly took ownership of it all.

In 2007, Jean Blaise reminds those years and says: "A few cities have used, as Nantes did, culture as a lever for economic development. It's a real challenge that is today a success. We bet on the international and the unusual, with symptomatic events symbolizing the mutation of the city like "les Allumés". National press followed with enthusiasm. Within a year, the image of the town as radically changed. Major artists like "Royal de Luxe", choreographer like Claude Brumachon, have arrived. "Play it like in Nantes" means that demanding and singular artistic forms are shown but always, with the idea to please the public. For instance, we gave the opportunity to provide in Nantes, many occasions to rediscover their industrial heritage, but also new settlements like the "Ile de Nantes". Culture as well as economic development and social binding, but also a way to conquer unexpected

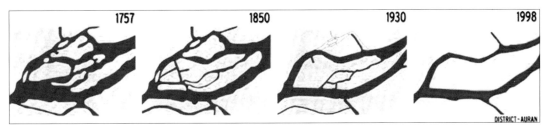

Fig. 2 Nantes islands in the Histiry

places. In the "Lieu unique", we welcome no less than 490.000 visitors every year, without great festival, just with a collection of tiny events. We also shelter artists. With the biennale "Estuaire 2007", we want to achieve a territorial project, creating a large single entity from Nantes to Saint-Nazaire. This territory is already an economic reality. However, the inhabitants not yet appropriated the banks of the Loire. The estuary, home place of major industries that have contributed to the economic development, will now be given to pedestrians ".

Fig. 3 Before the filling of the Erdre

Urban planning vs. landscape planning.

In December 1992, architect Dominique Perrault conducted an exploratory study and drawn major principles of intervention for the city of Nantes (Perrault & Grether, décembre 1992). In his chapter called "Method", the architect considers unrealistic to establish a general plan for the development of a vast territory. Interventions must register "different and concurrent" topics without hierarchy. He suggested to promote new installations, connections, and participation. Before 2000, Nantes didn't develop by strata, by layer, with a new sheet covering an old one. Each period had conquest a new territory, developed it or abandoned it. This resulted in a very strong sectorization with virtually sealed borders and striking contrasts of scale. As a consequence, industrial sites, located in the inner center on the island of Nantes, turned to lost areas, dangerous zones. It was like a dead spot in the very center of the city.

Fig. 4 Before renovation, flow of cars

Fig. 5 today's situation

Let's understand the situation. In the western part of the island, huge industries left big empty spaces. In the center part, old housing was held by poor families and in the eastern part, new housing seemed so desperate that the only wish there, was to leave as soon as possible. The island was cut into three parts by a large and fast motorway that brought people living in the south part of the city directly in the center. Perrault's answer was too

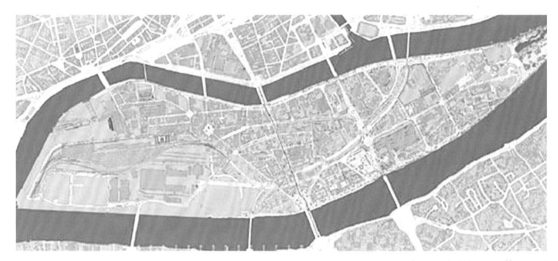

Fig. 6 Plan-guide (Chemetoff , 2010)

vague about what to do with this bulky industrial past and out of date Urbanism. We've seen that cinema showed Nantes as a city of cranes. For the inhabitants, the image of their own city was more blurry. A large investigation has been made in order to determine the way to the Island of Nantes was perceived (TMO-Régions, 10/1998). As a matter of surprise, a great majority didn't know that the Island is one island. A lot of people thought that it was still a collection of small islands as Neighborhoods' names suggest it. In the enquiries, somebody said that he cannot walk the dog round the island, so it can't be true. The eastern part of the Island is called Beaulieu. It's an important dwelling area. For a three quarter of the inhabitants, it was a transitory place and 70 % of them were moving away within a period of six months. As often advocate architects, offices were mixed with housing with a result of a feeling of not being in a private area.

As we can see, at the dawn of 2000, Nantes was dealing with a serious urban problem. As we saw in the corpus of films, the main cultural heritage was an industrial site: the harbor, which was, at that moment, closed. The knowledge, the identity, even the landscape of cranes, ships and workshops was lost, wiped away by the mid-eighties crisis. In 1999, the city of Nantes launched another huge competition to rethink the former harbor area and all districts now known as the "Ile de Nantes". Despite recommendations from economists urging to follow the Bilbao example, the chosen solution preserved a "low profile" approach with slow improvements rather than building a huge complex to be a symbol of the city. The Guggenheim Museum Bilbao, designed by Canadian-American architect Frank Gehry was inaugurated on October 1997. Since then, industrial reconversion, cultural developments bound with economic improvements seemed to have a magic recipe: a huge, spectacular building, a nice area around to spend money and a great names from the world of art and creation. This way of thinking has been rejected in Nantes, for philosophical reasons, but also for economic considerations: the risk of paying an expensive building with no certainty on economic sustainability was unacceptable in the context of job loss. The winning team, led by Alexandre Chemetoff, suggested, in a first draft, that he would work as an archeologist searching for hidden tracks of the recent past. His design preserved the ancient industrial buildings as well as local knowledge.

For this competition, three teams of architects were selected: Labfac, Bruno Fortier and Chemettof. Labfac designed a green sinusoid and Bruno Fortier a green promenade. "We must give to the isle of Nantes a strong image, a contemporary image, at the metropolitan level. " said Fortier. He thought of building a vast museum, reflecting what was done in Bilbao, giving the city a major attraction center within a green area.

Alexander Chemetoff went a totally different way. In his project, the island is divided between public and private spaces and shared between the "River Island" and the "Sea Island" areas. Chemettoff submitted as "game rule" through a plan"guide" that will allow stakeholders, public or private, understand and appropriating the overall project. The main problem in Nantes, and especially in this area is that the historical value of the urban spaces and buildings are not obvious.

No specific architecture, no spectacular constructions that would need to be preserved. Long discussions were held between the architect and local associations on what to keep and what to transform. In the end, everybody went to the conclusion to see in this area the ground of intangible heritage composed of knowledge, small histories and a specific way of life. The challenge was to convert it into urban design. First, Alexander Chemettof worked on the visual aspect of his drawings. He took the look and feel of city maps that everybody can find at the tourist office. By this, one can have the belief that the project is already done. Then, his main concept was division of spaces. The industrial area is characterized by vast plots that are not urban scaled. He redrew streets, places, esplanades. For him, no matter those spaces would become, it's just a question of the private / public area. Chemettof then worked as an archeologist, tracking all tangible remains of the recent past, to preserve them. He, of course, also protected the cranes as symbolic landmarks of the former harbor. Some industrial buildings are chosen to be renovated. The structure is preserved, the rest is destroyed. New parts are designed to be easily removed. Nothing should last.

The first idea was to develop a kind of business incubators for biotech. The concept was in the air and even if it wasn't rooted to anything in this area, it sounded nice for reborn of the industry. The problem was the loss of the intangible heritage and cutting the roots from the previous activities. Meanwhile, the great popular success from the big parades of the giants, led Chemettof to think of a new kind of continuity. Workers from the harbor could carry on working, in the same workshops, doing the same kind of activities but on a completely different purpose. Instead of building ships, they could construct a giant elephant.

A Giant wooden Elephant and Giant Little Girl.

In 2005 the most impressive image of all was created with the arrival of the giant puppets: the Little Girl and the sultan Prince and his Elephant. It is difficult to imagine how the most amazing feeling could be conjured up by a giant puppet, but seeing an Elephant, as tall as the tallest building in the city, walk through the narrow streets amongst stunned crowds is something that cannot be forgotten. Its success was so immense, the love of the city residents so deep for these puppets and their creators, that is has completely rejuvenated the city.

Following the regeneration of Bilbao, with its iconic museum, it was possible to imagine that culture could add value to a city looking for a second wind. In the early years of the new century the growing use of the internet and computers has brought prosperity and economic growth to the city of Nantes. But it also turned out that street theater, and particularly the wooden Elephant, has encouraged more people to visit Nantes, some even to stay, and the residents to love their city. This in turn has improved the economy with the influx of new talents.

During the mid-2000's, the theatrical production company "La Machine", split from the "Royal de Luxe" and built a new Elephant which became the emblem of Nantes. This Elephant, designed not as a piece of art but as a walking building, totally changed the way of urban thinking. In a way, the Bilbao effect came not from a huge iconic museum, but from a giant wooden Elephant, dragging tourists and residents to a part of Nantes that was, not long ago, seen as a "no go" area. After the elephant, artist and designer François Delarozière conceived many mechanical creatures, some for Nantes, some for parades, and some for other cities.

It's always surprising, to see such an urban project and believe that cultural and industrial heritage could be worked that way. Heritage can be seen as buildings, and Nantes kept the main landmarks to recall the past history. Heritage is also intangible like folk traditions, food and so on. In Nantes, heritage, finally, is not the exact preservation of what was there, but how it makes possible to write a new story, without destroying what was there before, but by overcoming the old ways. The success created perplexity and interrogations. One spoke about "the Nantes process", and studies were launched to dissect this process. But nothing was written or formalized before. What is being implemented at the urban level, is also at the architectural level, from sketch to completion, each small box comes to take his place on the layout. Chemetoff said he left a blur document on which he gradually focusses. The more the image is sharp, the more one sees details, better understands the nuances. It is a powerful concept, that of 'living logic'. Chemettof said: "all work on the city requires you to accept a part in uncertainty, it's the same when you plant a tree. You choose a location, you choose an essence, but you don't know exactly how it will grow. A city is a tree, it is alive" .

In ten years, the city of Nantes has radically changed its image. To achieve that, it needed a lot of new images to be created. In the early 21st century, the new paradigm for a city, is to be able to generate a tremendous amount of visuals that would be spread through the web. When Jean-Marc Ayrault has been elected as Mayor of Nantes, the city was cowering from its past that was well identified by cinema: a skyline of motionless cranes, still as the city was. But then came the street theater pushing people outside their home, bringing them new dreams, new visions. By opening a competition on this wide urban wasteland, Jean-Marc Ayrault understood that a link could be made between urban design and street theatre, a new direction could be followed.

The inhabitants of Nantes become familiar with bold shapes, modern architectures and even adopt a wooden elephant around which they flock in mass on Sundays and public holidays. Have we forgotten the cranes, certainly a bit, they are heritage, they denote the harbor from the past or a Steampunk image accompanying the mechanical creatures. The island of Nantes took its autonomy and is no longer the bulky part of the city, but rather the opposite, it becomes its motor, its inspiration. Ten years and one struggles to imagine that this place could be otherwise, old images become blurred, fade, one could hardly remember what was there before the elephant station, before the renewal. The city of cranes, Nantes became a city of the elephant, before industrial city, it became a city of culture.

Is this experience suitable for other cities? Can it be applied elsewhere in France, Europe or China? Bordeaux, Nantes' great rival, moved in the same way, developing what we could now call a narrative urban design. Close to Nantes, la Roche sur Yon called Chemettof to repeat his "method" on a very controversial site. Chemettof asked François Delarozière, the creator of the elephant to imagine a story that would be told all along the works. Delarosiere designed mechanical creatures, just as the elephant that the public can operate. This fantastic world of mechanical animals travels round the world. It started with a giant spider in Liverpool in 2008, the Aeroflorale in 2010, in Dessau and last year Long Ma Jing Shen – the spirit of the Horse-Dragon, in Beijing, China. We can imagine that those fantastic creatures, as they did in Nantes, could help to connect past history, heritage issues and contemporary questions on development, sustainability and well-being.

As a matter of fact, the main question becomes "what is heritage?" In Europe, the main answer has been to freeze things and turn them into a dead object state. Ancient cities transformed into open air museums as well as former industrial sites. This model has been massively exported to the rest of the world. Problem is, in many countries, this excluded people from places they used to live. The next idea was also to protect the intangible. Now, UNESCO / ICOMOS is looking for new ways to bringing back life, art, creation, inventiveness into places that have to be protected for what they hold, but also for what they represent and for the feeling of wellness everyone can have when visiting them.

Fig. 7 The first elephant

Fig. 8 The new elephant

Fig. 9 The new dragon

Art has become a consideration in city development. Just like Montreal, the beating heart of the city of Nantes is now the "Quartier de la Création", meaning the "Area of Creativity". Schools of art, architecture, design, fashion, dance and cinema, located there, attract engineers, scientists of all kinds, philosophers ...The word "Quartier" can also be understood as "village". The western part of the Island of Nantes would be a village dedicated to creation. That's a very interesting way to rebuild a story on the remains of the past, to start a new era.

Bibliography

Alexander, C. (1978). A Pattern Language: Towns, Buildings, Construction. OUP USA.

Aumont, Y., & Daguin, A.-P. (1998). Les Lumières de la ville, Un siècle de cinéma à Nantes. L'Atalante.

Barrau, F., & Wester, P. (1998, 05). chantiers artisanaux et patrimoine maritime : histoires passionnées. Nantes Passion, pp. 32-36.

Chemetoff , A. (2010). Le plan-guide. Archibooks.

Collectif. (2004). Nos amis les Français : guide pratique à l'usage des GI's en France 1944-1945. Le Cherche Midi.

Europe's secret capitals, The Last Best Place In Europe? (2004). Time magazine, Vol. 164, No. 8, August 30.

Île de Nantes : une ville se construit sous nos yeux, Alexandre Chemetoff ou la logique du vivant. (2007). Place publique #4, pp. 36-39.

Jeunet, J.-P. (Director). (2001). Le Fabuleux Destin d'Amélie Poulain [Motion Picture].

Perrault, D., & Grether, F. (décembre 1992). Au cœur du Grand Nantes, l'Ile de Nantes, Etude exploratoire pour l'aménagement de l'Ile de Nantes.

Saranga, K. (1993, 10 07). Quand Nantes S'allume. L'Express Spécial Nantes.

TMO-Régions. (10/1998). Ile de Nantes, Etude qualitative auprès des habitants.

UNESCO » Culture » Intangible Heritage. (n.d.). Retrieved from UNESCO: http://www.unesco.org/culture/ich/

吴晓淇
Wu Xiaoqi

教育背景
1986年　无锡轻工业学院（现江南大学）工业设计系本科毕业
1999年　中国美术学院 环境艺术系硕士研究生毕业
1993年至今　中国美术学院讲师、副教授、教授

科研课题
主持2010上海世博会浙江馆建设项目（省重大文化工程项目）

论文教材
《环艺设计思维快速表达》华中科技大学出版社2009.5.1
《产品与居住·公共家具与空间》中国建筑工业出版社

设计项目
马鞍山市采石古镇北区建筑设计
张家界国际旅游商业城景观设计
浙江省义乌市幸福里电子商务园室内及景观设计
浙江浙能六横电厂工程厂前区/厂区景观绿化设计
浙江舟山科技园/海洋生态环境检测站
马鞍山市秀山新区核心区城市设计及色彩规划、城市家具设计
太原市图书馆改扩建工程室内精装修设计

Education

1986 Graduated from Wuxi Institute of Light Industry (now Jiangnan University), Majored in Industrial Design
1999 Got Master's degree in China Academy of Art, majored in Environmental Art
1993- School of Art and Architecture, China Academy of Art

Research projects

2007.12 - 2011.12 Presided over the Zhejiang Pavilion Construction Project of the 2010 Shanghai World Expo (major provincial cultural project) of Zhejiang Provincial People's Government,

Recent Paper, Textbook and Lecture

"Quick Expression of Environmental Art Design Thinking" (excellent course on art design for institutions of higher learning), Huazhong University of Science and Technology Press, 2009. 5. 1
"Product and Residence·Public Furniture and Space", China Building Industry Press, 2014. 7
Representative Projects of Recent Design
Architectural Design of Ma'anshan Quarrying Town North Area
Landscape Design of Zhangjiajie International Tourism Commerce City
Interior and Landscape Design ofHappy LaneE-commerce Park in Yiwu, Zhejiang
LandscapeGreening Design of the factory of Liuheng Power Engineering Plant in Zhejiang
Zhoushan Science Park / Marine Ecological Environment Monitoring Station
Urban Design, Color Planning and City Facility Design of core area of Xiushan New Area in Ma'anshan
Interior Decoration Design of the Expansion Project of Taiyuan Library

历史城镇再生的一种方法
——马鞍山采石古镇城市更新

吴晓淇

【摘要】城市更新如同生命的有机体，伴随着城市的发展。如何在城市更新的同时良好保存城市渐变的印记，是建筑历史遗产保护的重要命题。人类文明的历史前行，一次次的变革，其文明的印记被城市这个显性的"历史容器"不断存载、积累。如同柯林·罗所谈的"城市拼贴"。"城市的拼贴"和生活方式的本土性成为了城市延存的重要历史价值，亦是今日成为文化旅游目的地的重要元素。中国改革开放三十多年，经济的飞速发展造就了城市的更新日异。但同时我们亦看到了一个残酷的现实：由于在城市变革中轻视了对城市历史记忆的保存，致使今天许多历史城市消亡变成"千城一面"。本文借助于安徽省马鞍山市采石古镇更新案例，探讨了古镇再生的建筑规划方法。希翼通过案例的研讨，探索历史城镇良性发展的路径。我们认为，城市如同生命伴随着记忆而生长，但仅仅把城镇发展的历史印记作为标本保护起来，一味赚取旅游"票金"，而忽视了与城镇共生的原住民，忽视了城镇自身的发展，是城市发展的又一种"自杀"。只有尊重历史发展的规律，以可持续发展的眼光，植入今日的生活方式与审美特征，城市才有可能"活"得自然，科学。在城市生长的同时，留存其生命的印记，如同习总书记所说的"看得见山，望得见水，忘不了乡愁"。

[Abstract] Urban renewal gets along with urban development like a living organism. It is an important proposition for the protection of historical architectural heritage that how to keep the mark of gradual urban change with the urban renewal. The marks of human civilization in the historical progressing and repeated revolutions are stored and accumulated constantly by cities, a dominant "historical container", like the "Collage City" written by Colin Rowe. The "Collage City" and the aboriginality of life style become the important historical value for continuing existence of cites, and also are the important elements for becoming a cultural tourism destination at present. China has reformed and opened up for more than thirty years, and the rapid development of economy has caused cities to be renewed and changed everyday. However, at the same time, a cruel reality appears: because we neglected the preservation of historical memories of cities in the urban revolution, many historic cities have become to have same urban features at present. In this article, we take the renewal of Caishi Town in Ma'anshan City, Anhui Province as an example to discuss the architectural planning methods for renewal of ancient towns, and we hope to find out a road for the sound progress of historic cities and towns through the research and discussion of cases. We think cities will grow along with memories like the life. However, if we only protect the historic marks in the urban development as samples, earn the money of tourism tickets blindly, and ignore the original residents symbiotic with cities and towns and the self-development of cities and towns, it is a kind of "self-destruction" for urban development.

一、绪论

城镇的"建设"与"保护"始终是一对矛盾。一方面由于经济的发展，城镇人口的增长，交通发展的压力，使得原有的城市空间结构已不能适应日益发展的城市需求；另一方面，由于忽视了发展中对原有城镇历史文化和肌理的保护和尊重，原有的城镇特色在建设高歌猛进的浪潮中逐渐消失，城镇的建设与发展造成了城镇的"千城一面"。如浙江德清的新市古镇，这个有一千二百余年的历史古镇，位于杭嘉湖平原的太湖之滨，河网纵横，典型的江南水镇空间结构，可由于常年建设的开发，现如今只剩下了三湾一塘，虽然这样的风貌亦在2008年被列入了国家历史文化名镇。浙江绍兴的安昌也是一个千年古镇，现在亦只留下了一条不足千米的水街。改革开放三十年国家经济获得了翻天覆地的变革，在这同时，我们亦付出了大量历史文化遗产消失的巨大代价。今天的政府逐渐意识到了历史文化的重要价值，城镇文化的保护，受到了地方政府的日益关注。在国家、地方二级领导关注下，"古镇复兴"成为了今天城镇建设又一"时髦"的名词，甚至成为了"运动"。但这种"运动"基本停留在了对旅游的狂热追崇。"古镇复兴"基本陷入了两种状态：一为静态的标本式保护，把原来的古镇原住民赶光，重新仿古建设，圈起来卖门票变成了一个旅游景点，这类案例不胜枚举。二为"假古董"和"再造古城"。山东的台儿庄，河南开封的上河图……在"古镇复兴"中是一味追求旅游效应而把原住处民驱走变成一座展览城，道具城还是真正研究城镇原有的地缘肌理和历史文化价值，使城镇和居民共生，同时在今天的审美和生活方式下可持续生存并兼有旅游价值，是今日"古镇复兴"中的重要命题。刘易斯·芒福德在《城市发展史》中有这样一段话："城市从其起源时代开始便是一种特殊的构造，它专门用来储存并流传人类的文明的成果；这种构造致密而紧凑，足以用最小的空间容纳最多的设施；同时又能扩大自身的结构，以适应不断变化的需求和社会发展更加复合的形式，从而保存不断积累起来的社会遗产。"值得我们深思。

二、历史城镇的定义及价值

历史城镇概念源于1933年8月国际现代建筑学会在《雅典宪章》中所提出的"对历史价值的建筑和街区，均应妥为保存，不可加以破坏"。1987年由国际古迹遗址理事会在华盛顿通过的《保护历史城镇与城区宪章》中提出"历史城区"的概念，将其定义为"不论大小，包括城市、镇、历史中心区和居住区，也包括其自然和人造的环境……它们可以作为历史的见证，而且还体现了城镇传统文化的价值"。同时还列举了历史街区中应该保护的内容是：地段和街道的格局和空间形式；建筑物和绿化、旷地的空间关系；历史性建筑的内外面貌，包括体量、形式、建筑风格、材料、建筑装饰等地段与周围环境的关系，包括自然和人工环境的关系，地段的历史功能和作用。我国在1986年国务院公布第二批国家级历史文化名城时提出了"历史街区"的概念："作为历史文化名城，不仅要看城市的历史，及其保存的文物古迹，还要看其现状格局和风貌是否保留着历史特色，并具有一定的代表城市传统风貌的街区。2002年10月修订后的《中华人民共和国文物保护法》正式将历史街区列入不可移动文物范畴，具体规定为"保存文物特别丰富并且具有重大历史价值或者革命意义的城镇、街道、村庄，并由省、自治区、直辖市人民政府核定公布为历史文化街区、村镇，并报国务院备案"。1994年面对东西方对文化遗产定义与保护的观点不同在日本奈良举办"与世界遗产公约相关的奈良真实性会议"。在会议上45位代表提出了《奈良真实性文件》，强调"文化多样性和遗产多样性"。在此基础上，文件在"价值与真实性"部分大胆地提出，真实性不能基于固定的标准来评判，反之"出于对所有文化的尊重，必须在相关文化背景之下来对遗产项目加以考虑和评判"。《奈良真实性文件》孕育于1964年《威尼斯宪章》之精神，并以此为基础加以延伸，以响应当代世界文化遗产关注与利益范围的不断拓展。我们从《奈良真实性文件》中可看到历史城镇作为文化载体的重要传承者——原住民的重要性。历史城镇不是简单地标本式的机械保护，而是在跨学科研究条件下对历史城镇、街区、建筑科学地可持续发展的保护。

三、历史城镇的保护和利用

从世界范围而言,工业革命之前,科学技术落后,经济发展迟缓加之不断的战争与瘟疫,历史城镇的破坏甚小。工业革命催生了科学技术的迅猛发展,带来了经济的高速成长。发展导致了原有的城镇结构对新需求的不相适应。许多城镇扩张成为城市,城市街区与道路极速扩张。在扩张中,许多历史文化城镇遭到了毁灭性打击。如何能在经济发展中人类文明和科学技术得到高速进步又能良好保护城镇发展的历史脉络,西方人首先进行了城市实践。1859年奥斯曼改建巴黎的城市建设规定中,对道路宽度从12米、18米到20米及20米以上,建筑高度最多不得超过20米,即除了底层以外上面最多建造5层房屋;到1884年开始结合道路的宽度规定建筑檐口以上部分的形状及高度,例如,道路宽度为12米,则新建筑檐口以上高度为½的道路宽度;1902年法规又做了进一步的修改,虽然建筑从地面到檐口的最大高度仍然为20米,但新的规定使新建筑檐口以上的高度大大增加,可建造两到三层,而不是过去的一层。1967年巴黎的城市规划指导中,建筑控制的方法有了新的改变,城市的尺度不再由道路确定,而是根据土地利用系数cos进行控制,整个巴黎平均为3。20世纪70年代兴起的城市历史文化保护运动中。1974年巴黎又通过一个修改的法案。新法规力图防止原有城市景观在新一轮的城市建设中被破坏,制定了更严格的控制规范,在城市历史中心区和某些有低层特点的外围地区,新建筑的檐口高度不得超过25米,中心以外地区为31米,这低于1967年的规定,并且新法规要求沿街建筑必须压建筑红线建造,某些允许退建筑红线建造的建筑其高度不得增加,以确保城市景观环境的整体和谐,科学严密的城市法规保护了巴黎的历史文化风貌,但同时并未限制巴黎的城市文明进步,时尚之都早已成为巴黎的代名词。在城镇历史文化保护和利用上,中西方由于民族文化特质的差异,导致对待历史和历史建筑的态度存在根深蒂固的区别,中国数千年的文化形成了一个超稳定结构,不变的宇宙观,微变的政治制度,不变的伦理信条使中国文化不间断地延续下来。而另一方面,历史城镇的保存亦在世纪朝代的更迭中在"新桃换旧符"的观念中丧失殆尽,只留下少数的历史建筑,西方文化从公元前1世纪就产生的民主政治,并以科学、革命、创新成为了人类前进的火车头,从有到实体。这种重实体的观点使得西方人很早就对自我文明的发展极为尊重。历史文化遗产保护的概念早在公元2世纪就已形成,古罗马著名旅行家波萨内斯曾记述,希腊奥林匹亚的赫拉神庙,最初的木柱大部分已被石柱替代;现代考古表明,赫拉神庙同一柱廊上的石柱分别属于较晚期的不同时代的艺术风格。历史文化遗产的利用,中西亦存在巨大的不同。西方在历史文化遗产的保护上,除了对历史文化遗产认真研究制定严格保护法规限制破坏外,还对原历史文化遗产科学利用并非修旧如旧,而是保留历史痕迹在现代技术条件下进行重新改造。如著名建筑师诺曼·福斯特改建的德国国会大厦。而中国则较注重对历史遗产的文本式保护,所谓"修旧如旧"。这种保护

图1

方法的结果，国家要投入大量的资金维护、整修，历史文化遗产只有瞻仰的价值而没有利用发展的价值。近年来政府逐渐意识到了如何以可持续的发展眼光，对历史文化遗产进行生态性的文化保护。

四、历史城镇再生之实践——以马鞍山采石古镇城市更新为例

2010年笔者受马鞍山滨江新区建设指挥部委托进行了马鞍山采石古镇的城市更新设计实践。该项目的提出建立在安徽省政府提出的打造"皖江城市带"的经济发展规划目标基础上。采石古镇历史久远，可追溯至春秋，在东汉时由于其特殊的地理优势成为了兵家必争之地，是古战场的发生地。唐代著名诗仙李白在这里追月投江，是李白的终老之地；今日中华文化传播的最早读本，《千字文》诞生于此。更有伯牙与子期的"知音之交"的历史传说。深远的历史源流印证了采石古镇重要的历史文化价值。近代，中国草书大圣林散之由于钟情采石古镇的青丽山水及对诗仙李白的敬仰长眠于此。可和中国许多城镇一样，今日采石古镇已是满目疮痍，面目全非（图1）。为了重新追回采石古镇的历史情缘同时亦为了与现有的，与采石古镇一河之隔的国家名胜风

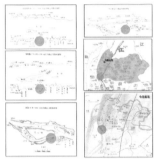

图2

景区——采石矶共同打造为国家5A景区，马鞍山政府委托滨江新区对采石古镇进行城市再生。接受滨江区任务后，笔者团队首先思考的是确立目标，并在目标指引下科学地开展设计工作，从而能真正把握采石古镇特色，重新找回历史情缘，在今天人类审美和现代生活方式上创新，可持续发展，在建设古镇的同时为马鞍山乃至长三角提供一个良好的文化旅游目的地。为此，我们在设计之初投入了浩瀚的前期调研工作。

1.古镇历史沿革及价值

据史料记载，采石镇大约在东晋南朝时期由于其直面长江，是当时都城建康（今南京）的一个重要卫城，拱卫首都的外围防线。据清嘉靖《郑开阳杂著》，为避长江水患，采石镇居民择高地而居，清光绪二十二年（1896年）长江中的江心洲名始现。乾隆二十三（1758年）年江心洲还称鲫鱼洲，此时的采石镇因鲫鱼洲阻挡，水势减缓，民居聚落应该扩展很大，沿锁犀河之半月布局基本形成。加上防水治水技术水平的发展，采石集镇已是繁盛之貌。民国时期，采石镇分南北两镇，北镇辖市北头，珍珠桥通津场、江口桥下街、中街，南镇辖上街、成民场、马巷、进士场、菜巷、敦仁里、河西。1952年4月29日，当涂县下区名以地命名，采石区辖采石、汤阳、宝庆、金议、三旺、金太、普集、西江、尚锦10乡县。1958年3月，采石镇为马鞍山市区三镇之一。10月29日，改镇为区，又独立马鞍山市区。1962年恢复区制。1964年该采石区为采石街道。1976年并入雨山至今。（见图2）

采石矶历史悠久，巍峙江左，古称牛渚矶。据《舆地志》记载，古时矶下崖洞中有"金牛出渚"，故此得名。秦并六国后，置牛渚县，汉承秦制，改为牛渚岩，晋设牛渚镇，隋唐时此处即驻有重兵，又称牛渚戍（见图3）。由于地形险要，采石历史上就是军镇。旧志载：采石之险甲于东南，其地南接楚疆，北络吴会，扼三江之襟要，一向是群雄之夺的战略要地。自孙策渡江袭牛渚，开创东吴立国基地后，司马炎在吴俘孙皓。陈后主遣兵战采石，韩擒虎宵济在陈朝，曹彬挥师取南唐，以及宋吴采石之战，清军与太平军之战，

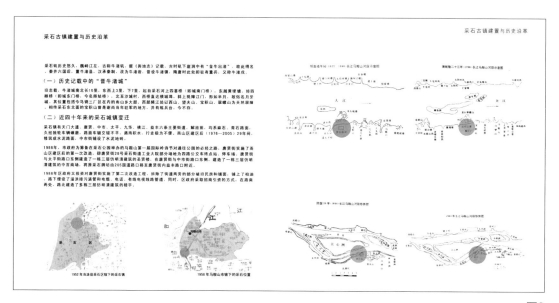

图3

都曾在这里反复争夺，南北交兵，发生过许多战争，历史上众多叱咤风云的人物，如周瑜、孙权、温峤、谢尚、陈霸先、黄巢、曹彬、虞允文、朱元璋、陈友谅、洪秀全、石达开都曾亲率士军，血战采石，使其成为江南著名的古战场。采石名人轶事众多，著名的有：楚霸王项羽、三国的朱然、东晋的谢尚、南朝齐人谢朓、唐代诗仙李白、白居易、南唐的樊若水、北宋郭祥正、南宋文天祥、清代黄仲则、清初萧云从。现代的郭沫若、林散之。更有南朝梁人周兴嗣在此撰写《千字文》成为中华文明的通识读本；采石亦建有佛、道二教寺庙。

2. 古镇的地缘肌理与特色

古镇位于马鞍山市区西南，其拥有陆路与水路二类交通，陆路东临天门大道（原205国道），西与采石矶隔溪相望，北有宝积山（荷包山）、小九华山依偎，南临采石河，西北部紧邻长江，自然山水条件优越；水路则为长江货运口岸。古镇呈半月状"怀抱"翠螺山（采石矶），其地貌中部隆起，南北二侧低缓，北高南低，形成特有的扇形结构（见图4）。以翠螺山为中心形成散射状的空间视域，形成古镇中横江、九华、中市、唐贤、太平、益丰六条街道的城市扇形结构（见图5）；确立了古镇以翠螺山为中心从属于山水这自然渊源，从属于特有文化发祥地的因果关系。采石古镇地域特色是自然的山水到古镇格局、到园林体系到城市布局的过度关系。由此呈现了九重的空间扇形结构。（见图6）

第一重：南北走向的长江沿岸。马鞍山采石矶风景区位于长江南岸，许久之前就是长江中的一块五彩石。之后命名为"采石矶"，与长江的渊源密不可分；第二重：三台阁、翠螺山同心布局。采石矶为长江中突起的一块基石，呈东西走向横卧江中 第三重：环翠螺山自然休闲林带；第四重：弓形的锁犀河；第五重：沿锁犀河的环状空间；第七重：古镇由道路而切割的环状区域；第八重：围绕天门大道所形成的环状自然绿地；第九重：南北走向的天北大道。古镇东西向轴线丰富，北低南高，中部隆起由采石矶延展而成。

135

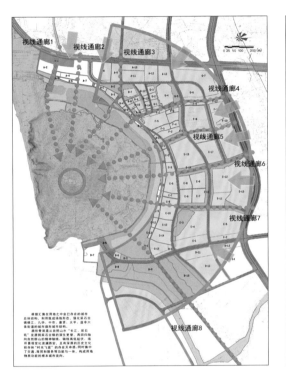
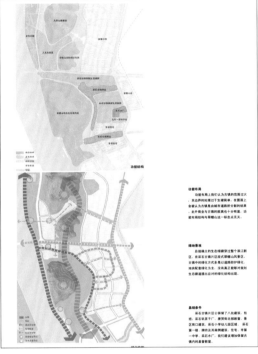
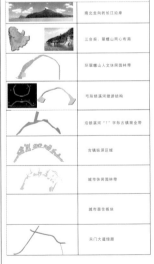

图4（左上）
图5（右上）
图6（下）

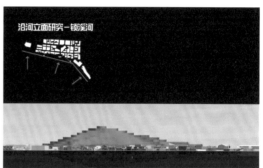

图7（左上），图9（右上）
图8（中）
图10（左下），图11（右下）

3.古镇的机遇与挑战

采石古镇历尽千年风霜，现今呈现的状态是居民聚居散乱，许多是居民自我违章随意搭建的棚屋。古镇仅有唐贤街较成集市，另尚有一所中学和一座基督教堂留存。一个悠久历史积淀深厚的古城如今变成了废城。2009年，马鞍山市政府根据安徽省提出打造"皖江城市带"的城市发展规划制定了"1255"城市发展战略。其中"1"主要任务优化主城区，城市西拓东进，滨江新区将成为主城区西拓东进的重要一翼，是城市发展的必然选择和当务之急。滨江新区总体结构可归纳为一轴六板块。一轴为九华山脉、荷包山、采石矶、采石河生态绿轴。六板块：都市休闲板块、文化旅游板块、拆迁安置板块、主题游乐板块、科技服务板块、都市生活板块。新区打造三大核心片区分别代表马鞍山的"过去、现在、未来"。采石古镇将成为"过去"的主题核心。重点体现传统古镇风情的人文与历史。采石古镇的城市更新成为了马鞍山市的重点任务，滨江新区的核心工作。有了机遇随之而来的是挑战：是简单重复国内现有大多古镇复建的模式：把原住民迁移，依照历史蓝本，仿制营建一些假古董，引进商业成为一个古意盎然的文化旅游"古镇"；还是认真研究、解析古镇深厚的历史底蕴，并站在今人的审美时尚和生活方式上探索古镇在"古"意上创"新"，面向未来打造一个生活与旅游共存，游乐与栖居共享，同时充分反映地缘肌理的长江风情市镇。这是我们进行古镇城市更新须明晰的目标。

4.古镇之再生设计

确定了城市更新的目标，设计即紧扣目标而展开分为：a.业态研究：在充分调研传统老字号江东特色商业街名号基础上，如何植入现代生活方式的商业业态，如咖啡、酒吧、时装精品等，使古镇不仅焕发古韵更能显现今日时尚之风的现代业态。b.文化研究：这是做得工作最浩繁的部分，必须查找大量史料文献，厘清采石现有的文化历史资源，并分类梳理在挖掘传统文化渊源的基础上，如何以现代创新方式呈现，如设立文化广场、博物馆、碑廊、诗院等，努力寻找适应今日现代生活状态的文化表现方式；为表现千字文的重要中华文化表征作用，以大地艺术的方式设计了千字文广场大街（见图7）。c.空间结构：在九重扇形空间结构指引下重新规划古镇街巷肌理，呈现采石矶地缘地理特质的城镇网络空间架构。d.建筑风格：这是设计中与甲方纠缠最多的问题，由于人们的惯性思维——谈安徽非徽居风貌不可。采石西据长江，长江从嘉陵江上游一路向东，在采石矶一折改为向南奔流，这里是当年楚霸王不愿过江的江东之地。因此其传统建筑风貌并非徽居风格，更多显现晋、唐之风。但一味地复古笔者团队认为是无出路的方法，因而力求传统出新，结合现代审美、技术、材料表现古镇新貌（见图8）。在经过会审通过笔者团队完成的控制性详细规划基础之上，滨江新区又委托笔者恩师——深圳大学建筑城规学院吴家骅先生率团队与笔者团队合作进行了北区的修建性详细规划设计。深入探讨了山水、建筑、空间的三者关系（见图9），尤其（恩师的睿智使设计更加传统出新，显现国际风范）深入检讨了建筑风格的形成。我们认为今日再造采石必须面向未来，照抄复制所谓的传统中国江南建筑特色，创造"中国江南味"的新江东建筑（见图10）。e.多样性、复杂性并存的街巷空间。街巷的多变，空间的转折，是构成古镇商业空间的重要之素，设计中研究街巷空间的异同，利用不同街巷尺度，营造丰富、有趣的商业集市氛围兼带现代审美趋味（见图11）。f.生态可持续。历史沧桑巨变，使古镇环境造成了重大破坏，如何尊重地域气候，保持地方物种的生存，又在建设中运用新型科学技术，生态养持，雨水收集处理，同时建材充分考虑节能之可持续发展理念，使古镇生态延续。g.研讨适合当地经济发展的商业模式结合商住转换，探究商业的本土性；研究如何利用地形高差营建生态、掩土型车库，良好疏解车行交通。确立交通方式，在研讨采镇住民与游客容量条件下，提出以步行交通为主，兼顾公共交通的交道处理办法。并满足住民人群的车行出行需求。

五、形、神、象，古镇再生之认知

　　中国传统绘画讲形似和神致，强调人的精神意象。历史建筑遗产的保护，尤其是一个现已不存但拥有深厚历史底蕴的古镇，如何能再生、复兴是一个值得深思的历史命题。仅仅从表象完全仿制，重建古镇最繁华历史时期的风貌，所谓再现，也可能是形似却失去了古镇的"气"，历史延存的"魂"，只会变成一个"假古董"，一个山寨的古镇剧场。虽然中国的国民好追异，怀旧之风今日可能会创造一定"戏剧性"的票房效果，可面对未来，古镇能否可持续生存？所谓"形""神""象"既要准确把握古镇的特征谓"形"，又要捕捉古镇的风韵曰"神"，这种风韵决非标本式的复制，而是依据其地缘肌理特色在久远深厚的历史文化土壤中焕发出来的古镇之气。更要聚焦于古镇的"象"。何谓古镇之"象"，老子说"大象无形"谓人生之至高境界。如能真正表现古镇之象，古镇再生就真正在反映古镇特色风貌之上表现出了发展的进步，持续的人类生存之象，这是我们为之不断苦苦追索的结果。由于种种原因，本设计只停留在了文本、图纸阶段并未实现。此亦是笔者需认真反思的：如何提高领导者的思想意识，真正为振兴文化，再生古镇寻找科学之路，是我们每一个建筑历史遗产保护工作者的时代责任。

参考书目

[1]《历史城市保护学导论—文化遗产和历史环境保护的一种整体性方法》张松著　上海科学技术出版社，2001年12月
[2]《城市发展中的历史文化保护对策》张凡著　东南大学出版社，2006年8月
[3]《THE CITY SHAPED: Urban Patterns and Meanings Through History》Original Drawing by Richard Tobias, Published by amangement with Thames&Hudson Ltd.London ©1991 Thames&HudsonLtd. London

周胤斌
Zhou Yinbin

2008年毕业于中国美术学院中德学院，获德国柏林艺术大学硕士学位
现任浙江西城工程设计有限公司总经理，高级工程师

设计项目

2010年　主持设计杭州"母亲河"中东河综合整治与保护开发工程，荣获钱江杯、西湖杯
2011年　主持设计南京市沿长江滨江带项目，获江苏省优秀设计
2012年　主持设计富阳市杭富沿江大道及富村山居黄公望建筑景观项目荣获钱江杯、西湖杯
2012年　主持参与设计超山南园、西园综合整治项目
2013年　主持设计超山海云洞建筑景观项目荣获钱江杯、西湖杯
2013年　主持设计苏州国学大讲堂老太庙建筑项目，获苏州市优秀勘察设计奖

2008 Graduated from CDK CAA, Master of UdK,
General manager of Zhejiang XIcheng Engineering & Design Co., Ltd.

Research projects

2010　the design of Hangzhou "Mother River" Zhongdong River comprehensive improvement and protective development was awarded Qianjiang Cup and the West Lake Cup.
2011　the design of Waterfront Zone along the Yangtze River in Nanjing was awarded Outstanding Design in Jiangsu.
2012　the design of architectural landscape of Hangzhou Fuyang Riverside Road and Yuan-dynasty painter Huang Gongwang's Fuchun Resort won Qianjiang Cup and the West Lake Cup.
2012　took part in the comprehensive improvement project of South Park and West Park of Chaoshan Hill.
2013　 the design of architectural landscape of Meiyun Cave of Chaoshan Hill won Qianjiang Cup and the West Lake Cup.
2013　 the architectural design of the Imperial Ancestral Temple, the Traditional Classics Grand Auditorium of Suzhou, won Excellent Survey & Design Award in Suzhou.

同筑共赢，重构美丽江南
——杭州拱墅胜利河古水街项目案

周胤斌

【摘要】建筑遗产保护的本质意义，即文化传承。如何用现代手法将遗产文化继承、延续及创新，将文化根植于人们的生活方式之中，从而使建筑遗产得以精神上的延续，将建筑遗产保护的意义升华。胜利河古水街集江南运河建筑、河道元素于一体，被赋予强烈的历史感、时代感，成为运河文化的载体。承载着大运河文化和现代的艺术，将永远展示在杭州的拱墅运河段。

[Abstract] The essence of architectural heritage protection is cultural heritage. It is very important that how to inherit, continue and innovate the heritage culture with modern techniques and to root the culture in people's life styles, so that the architectural heritage can be spiritually continued and the significance of architectural heritage protection can be sublimated. The Gushui Street on the bank of Shengli River integrated of elements of canal buildings of southern China and watercourse with strong historical sense and the sense of times has become a carrier of canal culture, bearing the culture of the Grand Canal and modern art, which will be shown in the canal section of Gongshu District, Hangzhou forever.

浙江历史悠久，历史建筑众多，建筑遗产保护成为了十分重要的一项课题，如今杭州有西湖和大运河两个世界文化遗产保护项目，如何在可持续发展的乡镇建设、旧城改造和历史文化遗产保护等方面表现出色，值得深思[1]。

一、杭州城市河道水系背景

杭州的环境优势就在于"水"，它集江、河、湖、海、溪于一城，是一座著名的"五水共导"的江南水城。千年运河从隋朝流淌至今，带给了杭州千百年来的繁华，并由此促成了杭州中国古都的地位，构成了杭州城市重要的历史文化遗产和文化景观。而杭州城市发展的历史，就是与京杭大运河的开拓、发展和繁荣紧密相关的。

1. 大运河奠定了杭州的城市发展

杭州与运河共成长。杭州城市的形成与声誉，是与大运河相伴而生长兴盛的。"杭州"之名，由河而生。历史上第一次出现"杭州"之名，就是在隋朝凿穿大运河之后，始废钱塘郡，设"杭州"。杭州城市格局沿河而建，江南名城借河而扬，杭州凭借着京杭大运河与北京南北贯通，才一直保持着江南名城的地位[2]。

随着"建经济强市，创文化名城"的发展战略的实施，大运河沿岸地带的再开发必然成为杭州城区开发建设的重要组成部分[3]。

2. 大运河成为世界文化遗产

中国大运河是中华民族两千多年历史的见证，是保护中国古代丰富文化的历史长廊。大运河凝聚着政治、经济、文化、水利、建筑、科学、教育等多个领域的价值与信息，是一条历史之河、文化之河。千百年来，在运河（杭州段）沿岸汇聚了丰富多彩的茶艺文化、饮食文化、桑蚕丝绸文化、地方戏曲、民间曲艺，积淀了古典园林、藏书楼阁、桥梁古塔，形成了运河沿线著名的"湖墅八景"等人文景观。所有这些，既是运河文化丰富内涵的体现，也是运河申报世界文化遗产的宝贵资源。中国大运河（杭州段）是运河目前保存较完整的区段之一，杭州在中国大运河申报世界文化遗产中正发挥着特殊而重要的作用[4]。

2002年杭州开始实施大运河（杭州段）综合整治和保护工程，明确了"还河于民、申报世界文化遗产、打造世界级旅游产品"三大目标。经过多年的整治，取得了显著的成效[5]。

2014年6月22日上午，在卡塔尔多哈召开的联合国教科文组织第38届世界遗产委员会会议审议通过中国大运河项目列入《世界遗产名录》，成为中国第32项世界文化遗产[6]。

大运河申遗成功，对我国来说，具有独特的历史意义和现实意义。首先，大运河"走向世界"，有助于世界了解中国，了解中国悠久灿烂的文化。其次，大运河的申遗成功，使世界遗产的保护理念在中国大运河沿线8个省（市）、1.7亿民众中得到广泛传播并深入人心，更有助于国民了解自己历史，激发民族自豪感，激励人们进一步改革开放与开拓进取，振兴中华。第三，有助于对大运河遗址的更有效保护[7]。

3. 胜利河——连通大运河与上塘河的重要枢纽

大运河与上塘河、苕溪、钱塘江为杭州市内四大水系。

上塘河是杭州历史上第一条人工河，相传最早由秦始皇开凿，一直是大运河进入杭州的唯一通道[8]。

钱塘江干流在杭州境内从建德到桐庐、富阳再到市区[9]。

胜利河是大关地区的"母亲河"，是连通上塘河和京杭大运河的重要枢纽。在上世纪90年代，胜利河曾经遭遇严重污染，河水发黑发臭。但是经过10多年的整治，清清的河水流入了世界文化遗产[10]。

二、运河文化背景

运河文化的一个突出表现，是外化为一种商业文化，正是因为运河的航运功能，使得运河两岸繁盛一时，呈现出一派繁华的商业景象。当然，也正是因为如此，不论是运河的航运功能，还是由之而来的沿岸商业繁盛，都要求因时而变，与时俱进，造成运河两岸景观变化迅速，历史文化古迹存留至今相对较少。

"运河文化"的形成和内涵可以归纳为：因水而兴运，缘运而聚商，倚商而成市，随市而显貌，貌以时迁，随时而变，并与时俱进；"运河文化"的本质是一种流动的文化、变迁的文化，是显示时代风貌、折射时代特征、并且追随时代潮流而进的文化。

1. 水陆枢纽少不了喧闹

上个世纪初，沪杭铁路未通前，杭州城里并不热闹，热闹的反而是临运河拱墅段的拱宸桥。这里既有拱宸桥轮船码头、火车站等客运枢纽，又有戏馆、茶园等娱乐场所。

运河拱宸桥曾沦为日租界，自开埠后，日本人不遗余力促其繁荣，何扬名曾在《杭州文史资料》第18辑中回忆，"凡可助市场发达之事，虽鄙贱亦不禁阻，故首倡公娼，以招徕游客。"

2. 运河茶园，不可不知的运河茶事

南宋以降，十里湖墅，水陆码头众多。挑夫、船民、渔民要歇脚，生意人要洽谈，名人雅士要品茗论道……沿着运河一路过来，以桥为节点，两岸原本就已茶馆密布，有着良好的基础。中日甲午战争后，杭州开埠，拱宸桥一带成为通商口岸，拱宸桥东岸茶馆的档次便也为其他地方所不能比肩。这些茶馆，当时叫茶园，不仅卖茶，还演戏，深受有钱人和戏班亲睐。

清光绪二十一年（1895），拱墅运河畔首家茶园——天仙茶园开张。后来，相继建起了荣华、景仙、丹桂、阳春等茶园。一时间，拱墅运河畔出现了"洋街两面沸笙歌，戏馆茶园逐渐多。国忌如今都不禁，日间弹唱夜开锣"的情景。

清末，拱墅运河畔不仅有中国人开的茶园，还有外国人开的茶园。1896年至1908年，经过12年时间的发展，拱宸桥地区由原先一片庐墓与桑田错杂的旷野之地变成了号称"小上海"的洋场，西方文化源源不断地输入杭州。

屹立的拱宸桥，是运河南端的丰碑；运河上倒映的拱宸桥，既是建筑和自然的景观，也是历史的倒影。从戏曲到影视，运河见证了中国戏曲史自清代以来在杭州的发展[11]。

3. 工业发达首屈一指

因运河的航运功能，建国以来大批现代工业在运河沿岸的崛起，运河两岸工业的发达，首屈一指。除了杭州第一棉纺织厂，还有浙江麻纺厂、杭丝联等。杭一棉，最多时曾有近万名工人。

4. 运河景观体系

古代时期运河主要是交通往来的通道，以旅为主，但也出现了将两岸景观作为欣赏对象来观赏的"游"的意味，只是并未作为固定的游览对象，也未刻意加以经营。史载"水牵卉服，陆控山夷，骈樯二十里,开肆三万室"，以及"邑屋华丽,盖十余万家，可谓盛矣"等[12]，其繁盛的城市景象已经引起了人们的观赏；北宋词人柳永也留下了著名的"东南形胜，三吴都会，钱塘自古繁华。烟柳画桥，风帘翠幕，参差十万人家"，以及"市列珠玑，户盈罗绮，竟豪奢"的词句；明清之际，运河沿岸还出现了以"湖墅八景"为代表的景观体系[13]。

三、杭州胜利河（上塘河—运河）沿线景观整治设计

项目设计范围为胜利河区段，东接上塘河，西联京杭大运河水系，北临规划霞湾路，河道全长约1.5公里。胜利河周边土地利用主要是居住用地以及文教用地。其北侧从西到东依次是富义仓、八丈井新村、大关社区，其南侧从西到东依次是霞湾公园、德胜小区、浙江省古建筑设计研究院、假山新村、杭州市国家安全局、德胜新村、杭州市德胜小学、杭州市交通职业高级中学（图1）。

1. 项目立项思考

1）本案重要性

①运河文化和西湖文化、南宋文化为杭州最大的三块历史文化，是具有申遗价值的珍贵历史文化。②杭城的南北两块文化比重不和谐，与杭城南块的西湖文化、南宋文化能"匹敌抗衡"的唯有运河文化。拱墅具备运河文化的"专利"特征。③目前在拱墅众多的运河文化亮点呈现的是散、稀、弱，有必要打造一座具

图1：胜利河现状平面图

图2 朱仁民创作北关夜市30米手稿

图3：朱仁民创作相关手稿

有独一无二的、代表性的文化精品作为运河文化的红线，串起拱墅众多的运河亮点。本案具有历史文化的深度和时尚现实的经济功能。从而为平衡杭城南北文化板块而起到积极的作用。④运河文化在杭城历史悠久，作为其载体的建筑、驳岸、桥梁、街道和民俗、生活、用具大量消亡，尽管有关部门作了极大的努力，但因为缺乏具体的结集型的深入挖掘、修复创作缺乏经典性的运河项目，无法在杭城树立运河形象。本案企图以"清明上河图"的手法，原汁原味恢复拱墅历史上"北关夜市"的当年景象（图2—3），还这段运河以历史真实的繁华景象。⑤本案以"水街"手法，打造具有代表性的运河经典片段，以"水街"为主要形象，表达运河历史原貌。抓住历史上这里为杭城茶楼酒肆最丰富的地段：鱼市、米市最为贸盛的场所，以"水街"上的旅游、经营、娱乐、购物为内容和杭城南面的河坊街形成南北一阴一阳两大历史文化场景。

2）区域优势

①胜利河作为杭州最重要的两大河流的纽带河，具有极其重要的价值和其本身意义。胜利河地处城市都市内中心的河道，两岸人口稠密，水路、陆路交通繁忙，其人流、市气极佳，在这里营造运河精品工程，便于游客和市民接近运河、了解运河、消费运河。②胜利河有拱墅地段所蕴有的丰实历史人文亮点，尽管这些人文景点分布散疏，但是随着运河、上塘河的水上游，水上交通的开辟，运河改造工程的深入，拱墅运河的文化点将串联成一条靓丽的珍珠项链。因为富义仓的存在，因为胜利河西头霞湾"三角洲"的存在，胜利河具备了拱墅运河段最富变化的水系形态。③作为拱墅杭城运河文化的最大承载区域，胜利河是上塘河和运河的"腰带式"连接，具有极高的旅游价值和交通功能。胜利河的特殊位置、形态和文脉价值，正好是拱墅运河文化体现的一个极好亮点。

3）本案主题构想

按目前的设计，将胜利河作为运河与上塘河的纽带河起着交通、过道的作用，利用河道的适宜宽度，建立水上自然景观通道的个性功能，并在整条河道中如同写文章一样进行由西向东的"虎头""熊腰""豹尾"，将河道分为二段式风格三个节点式处理。抓住节点，畅通二段，建成水上景观廊道怀古志新。

怀古：以霞湾公园和富义仓为基础的生态文脉。加强这一段河道两岸的古文化点缀。以船缆柱、旧驳岸、古埠头、古建筑、古桥梁为基础的古运河元素，不张扬、不扮饰，让船在历史的河道中通过，在体现此河道的交通功能之外，达到"历史水中走，小船水上游"。让游客在水中参与、购物、餐饮、娱乐等别开生面的运河夜游活动。建立"清明上河图"式的"运河风情图"实景，收购建设古代拱墅"井屋鳞次，烟火万家"的一个截面，"精帆卸泊，百货登市，篝火烛照如同白日"的"北关夜市"，恢复清代的钱塘十景之一。沿河两岸恢复民居临河而筑，吊脚、骑楼沿街长廊，既避风避雨又宜行宜离，与驳岸、码头、仓储相结合，形成半林半郭、半宏半高、半野半水休闲舒适的运河古代情致。拱墅运河边曾是杭州历史上茶楼酒肆最多的地方，戏曲、评书最热闹的地方（图4）。恢复这一段富有运河特色的场景，将与河坊街、梅家坞一样形成杭城特色旅游的区块，为拱墅的运河文化增添一大重要景点。

志新：在胜利河的东段北岸建立运河文化艺术会馆、邀请国内外一流的创意文化艺术家，进行作品长期的展示，为本案的专业文化档次提供强有力的保障。本设计中已具备中国近代绘画史中具有权威性的人物画长卷《天下粮仓》主题作品，以及20米长卷《拱墅河上图》均为运河文化传世之作，将其在这里永久展示，为胜利河乃至整个拱墅运河文化带来极高的艺术品位。同时，以艺术会馆为中心的自然公共绿地也为游客市民的文化休憩、学习交流带来极大的便利。

如是两段怀古志新将运河新旧文化统揽于上塘河与大运河之间（图5—6），和富义仓、霞湾公园组合成杭城中心区域最有分量的文化休闲地块。

3）本案建筑

本案的建筑分为怀旧、志新两大系统，桥梁、驳岸与河道遮阴蓬架作为建筑的连接构筑物（图7—10）。

①怀旧区

怀旧这段建筑均以恢复拱墅晚清期间河道上民居为主题的建筑群体。这些建筑在水路、陆路两侧外立面均以历史真实面貌为临本，原汁原味地反映历史真实景象。重要的观赏部位要精心刻画，并以收购和异地拆

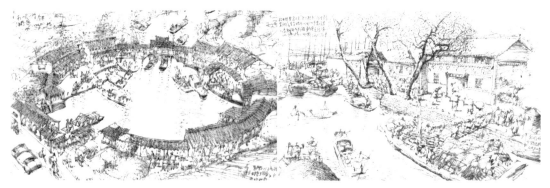

图4：朱仁民创作手稿

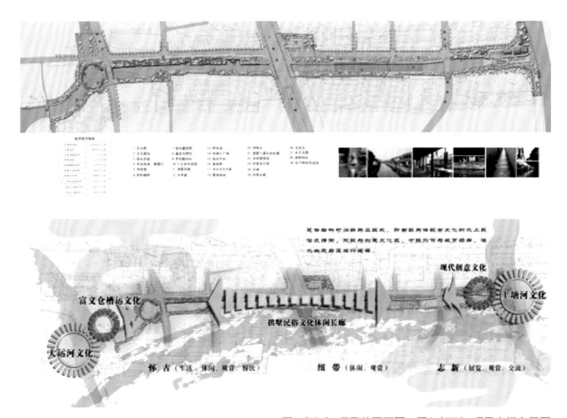

图5（上）项目总平面图，图6（下）项目空间布局图

造手段实施营造计划。临街临水的吊脚楼、河埠头、古石桥、驳坎侧石、打缆桩、花格窗、旧门古墙、民俗用具尽可能在民间和各市场专业收购，并和现代的高仿摹写相互穿插、渗透，使之修旧如旧，恢复运河的历史记忆。

②志新区

志新区建筑以当代中国四位创意性艺术家作品展示为中心，作为会馆式的小型观摩展览。作品以不脱离运河文化、水乡文化、水文化的现代创意型绘画、雕塑原作为主体，这些建筑力求在运河文化的元素中营造出现代的具有高端艺术特质的建筑作品，其建筑本身要求具备单件运河艺术品的高要求，并要与整个胜利河的古建筑、古桥、古驳岸进行个性和气质上风格统一、一气呵成。

③遮阴蓬架

遮阴蓬架应以原木或毛竹为主题造型，组合成现代的跨越式构架立于运河两侧，既考虑到游船交通的便利，又考虑到冬日夏日的遮阴挡阳，为水上旅游建立别开生面的运河通道。

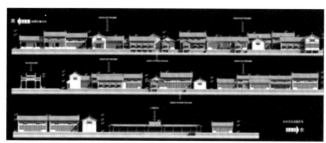

图7（左上）朱仁民创作手稿，图8（右上）建筑效果图
图9（左下）建筑设计模型，图10（右下）建筑实景图

④桥梁、埠头

本案中的桥梁根据旅游功能的需要和历史真实性的遵循，在石、木结构上进行仿古和艺术性处理，按每座桥梁的区位功能不同、设计不同风格的桥梁。在河道驳坎中选择合理的位置，建立仿古并具实用性功能的河埠头。埠头中可以雕塑形式展现古人河埠头生活的场景。

4）本案典型景观

①华光观渔

西段北岸通过高墙长廊连接富义仓，展现漕运文化，南岸则通过数艘渔船与栈桥来营造大兜鱼市与华光观渔的热闹场景与意境，使游船从运河一进入胜利河就能体味到湖墅鱼市、米市的喧哗繁荣。

②湖墅环楼（图11—12）

在胜利河与红建河交汇的三岔口，一组气势宏大、气氛热烈的环形建筑群楼与廊、桥一气呵成，成为整个河道的中心焦点与地标性建筑，也使得该河口西、南、北建筑连成一体。其间茶楼酒肆、评书曲艺、欢歌渔唱，热闹非凡。

③元帅庙会（图13—14）

拱墅运河边曾是杭州历史上茶楼酒肆最多的地方，戏曲、评书最热闹的地方。正对规划七号路处设置描绘牌坊、古戏台，再现评书、戏曲、相声、杂弹的运河戏曲庙会场景。

147

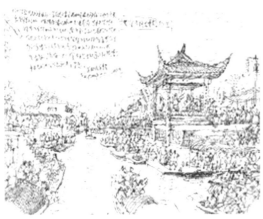 

图11（左上）湖墅环楼效果图，图12（右上）湖墅环楼效果图
图13（左下）古戏台朱仁民手稿，图14（右下）古戏台效果图

④八丈井村史图

八丈井村原是运河卖鱼桥外一个小村，随着城市扩张渐渐演变为城市居住区。设计拟在八丈井村村口位置即胜利河北岸、上塘路以西设景观节点，列运河农家小院，记村史变迁，形成富有历史文化的绿地小景。

四、结语

建筑遗产保护的本质意义，即文化传承。建筑遗产，是文化遗产大家庭中不可忽视的重要成员。他们见证着国家和民族的复兴之路，体现着建筑先辈为争取民族独立、国家富强、社会进步而前仆后继、自强不息的精神，凝聚着各个时期建筑师的崇高理想、信念、品德和情操，形象而直观，具有强烈的感召力。建筑遗产形成于过去，认识于现在，施惠于未来。文化遗产既是历史的，也是现实的，既是物质的，也是精神的，既有真实的感受，也有理性的思考[14]。而在当今时代，对建筑遗产的保护应除了本意上的对历史遗迹的修复和维护，还需要考虑如何用现代手法将遗产文化继承、延续及创新，还原依附于建筑本身不同时期的人类生活方式,巧思如何将依存于建筑古迹内的文化历史,更好地为人所知,更好地融入现代人的生活,将文化根植于人们的生活方式之中,从而使建筑遗产得以精神上的延续,将建筑遗产保护的意义升华。

在民间的个私建筑形态和现代商贸功能中取得吻合点是胜利河古水街的重要意义。取古水街民居建筑的强烈韵律感、色彩感,使它成为能观赏、能旅游、能营业、有交通、有消防、有暖通、有排污的现代旅游运河地标是古水街创作的终极目标。胜利河古水街集江南运河建筑、河道元素于一体,被赋予强烈的历史感、时代感,成为运河文化的载体。承载着大运河文化和现代的艺术,将永远展示在杭州的拱墅运河段。让民众将运河带回家,让环楼戏说,给子孙一个遐想,给城市一个记忆,为运河申遗尽献绵力。

参考文献

[1] 中国美术学院召开"建筑遗产保护"国际论坛.爱画网.院校新闻.
[2] 余小平.大运河(杭州段)水文化研究.江南文史.
[3] 胡晓聪 邹志荣.找寻逝去的足迹,还原两岸的风景——京杭大运河杭州主城区段沿线环境改造设计.
[4] 阮利平 史萌.杭州:流淌的运河文化.文化交流,2013年第七期.
[5] 陈伟 朱娟丽.大运河杭州段申报文化遗产景观提升规划设计.浙江建筑,第29卷,第5期.
[6] 大运河、丝绸之路申遗成功.地理教学,2014年,第15期.
[7] 柯继承.大运河申遗成功再思考.MODERN SUZHOU现代苏州,2014.8.
[8] 王国平总主编,陈钦周,杨卡特.杭州河道文明探寻.
[9] 唐玮洁.杭州水系地图 流淌在我们身边的河.青年时报.
[10] 杨骐恺.寻找胜利河的"前世今生".杭州日报.
[11] 任轩.拱宸亚,运河上的中国.
[12] 胡国枢,徐明.沟通江河,连接湖海——隋代大运河与杭州[A].周峰.隋唐名郡杭州(修订版)[C].杭州:浙江人民出版社,1997.42-56.
[13] (明)田汝成.西湖游览志[M].上海:上海古籍出版社,1998.233-242.
[14] 单霁翔.20世纪建筑遗产保护吹响"集合号"演讲稿.

贾科莫·皮拉兹·皮拉左里
Giacomo Piraz Pirazzoli

贾科莫·皮拉兹·皮拉左里，建筑师，曾以优异的成绩毕业于佛罗伦萨大学，并于罗马大学完成博士学位。随后皮拉左里在巴黎勒·柯布西耶基金会（Fondation Le Corbusier）从事研究活动，并就职于建筑师克里斯蒂安·德·波特赞姆巴克（Christian De Portzamparc）的巴黎办公室。从佛罗伦萨和米兰走出的皮拉左里，经常与一些知名同事合作，如Francesco Collotti、Paolo Zermani、Fabrizio Rossi Prodi等人，在意大利及其他国家拥有建筑、博物馆和展览馆等设计作品。现为佛罗伦萨大学建筑设计学院实验室主任。

就职机构
1997—2001年　　欧洲建筑师协会，委员会成员
2000年至今　　　佛罗伦萨大学，建筑设计学院副教授
2000—2010年　　建筑设计博士学院，委员会成员
2011—2014年　　iCad建筑设计国际课程，创始人兼董事长
2000—2003年　　意大利外交部，阿尔巴尼亚项目特遣小组成员
2002—2006年　　佛罗伦萨斯提波特博物馆和"阿道夫·卡尔米内"艺术研究
　　　　　　　　基金会美术学院，主席

主要经历
海外授课及访问学者经历包括苏黎世联邦理工大学、蒙彼利埃高等美术学院、EPOCA（布宜诺斯艾利斯）、埃里温国立建筑工程大学、哥伦比亚大学、马耳他大学、巴利亚多利德大学建筑高级技术学院、伊斯坦布尔技术大学、雪城大学、马德里建筑学院、阿勒旺大学（开罗）、马德里圣巴布罗大学、慕尼黑应用技术大学、奥本大学乡村工作室、圣路易斯华盛顿大学、库伯联盟学院等。同时，曾以"绿色城市规划——智慧城市"为题在TEDx演讲发表演讲。

Giacomo "Piraz" Pirazzoli, Architect PhD
director at CrossingLab.com - Department of Architecture, University of Florence
Giacomo "Piraz" Pirazzoli is an architect. Graduated with honours (Florence), he carried on research activity at Fondation Le Corbusier (Paris) after his PhD (Rome La Sapienza), while practicing at Christian De Portzamparc's office in Paris. A professional based in Florence and Milan, he designed several buildings, museums and exhibitions in Italy and abroad - often in collaboration with reputed colleagues Francesco Collotti, Paolo Zermani, Fabrizio Rossi Prodi and others.

Institutional duties include: European Architects Council, Brussels (1997-2001: Committee Member). University of Florence (2000 to present: Associate Professor on Architectural Design), Architectural Design PhD School (2000-2010: Committee Member), iCad_International Course on Architectural Design (2011-2014: Founder Chairman). Ministry of Foreign Affairs, Italy (2000-2003: Albania Project Task Force Member). Academy of Fine Arts, Florence and "Adolfo Carmine" Foundation for Art Studies (2002-2006: President). Stibbert Museum, Florence

"GreenUP — A Smart City"亦被称为建筑遗产保护理念

贾科莫·皮拉兹·皮拉左里

GreenUP — A Smart City是一个由Giacomo Pirazzoli教授牵头并协调的跨媒体研究项目。实际上这个项目是由CrossingLab.com的智库提出的。

GreenUP项目，作为一个高度灵活且经济实惠的垂直景观网络，重点是将文化及建筑遗产视为可扩展的领域。

GreenUP项目可适用于都市环境中通过已经存在的绿色植被进行改造整修，也可以依照新的规划从头进行设计。

GreenUP项目也可以将当代的建筑设计方法与历史建筑古迹保护和谐融合，从而实现可持续发展的人居环境。

目标

作为一个自下而上进行"全球化思考及本土化行动"的方法，GreenUP项目旨在提高环境质量，例如宜居性，可在减少空气污染的同时增加二氧化碳以达到平衡状态等等。

GreenUP项目同时也可考虑在绿色植被中种植蔬菜水果，这样可以保证食品安全供给，在社区内进行社会共享。

GreenUP项目的关键字

减少碳排放，大众货品，城市农业，回收利用，能源开发以及生物多样性

研究背景

GreenUP是一项由一百多位研究人员参与的全面的研究工作，其人员包括：学生，建筑师，植物学家，工程师，农业专家，城市规划师等。在过去的五年里，所有跨学科的研究工作都是在自愿参加的基础上进行的。

在"GreenUP—A smart City"概念框架下所提出的几个案例已与欧洲的合作机构进行了研究，同时我们也在寻求能与发展中国家开展这方面的合作。

2012年，研究团队在巴黎全球大城市管理国际会议中第一次向全世界呈现了这项多学科跨领域的研究预演。

在以"城市连接"为主题的TED会议成功举办之后（图1）：http://www.tedxhamburg.de/greenup-a-smart-city-giacomo-pirazzoli-at-tedxhamburg-urban-connectors, GreenUP项目的研究团体不仅在美国、澳大利亚亚太研究机构（科廷科技大学，珀斯）开展了研讨会及成果展示，他们还在欧盟及巴西气候变化大会（里约热内卢，2015）上进行了演说。

目前，基于一些国际合作项目的启动，GreenUP的后续研究也得到了充分展开。

跨媒体合作

GreenUP的跨媒体项目其中包括印刷书籍（Allemandi International，都林—伦敦—纽约 2013）及增加电子书的链接。

由电影导演Filippo Macelloni制作的侧面反映城市绿化的视频，及由艺术家Agnese Matteini创作的原稿已在https://vimeo.com/87993255展出。该网站www.CrossingLab.com 可提供预览。

GreenUP 在中国

出生于中国的任雪飞博士也非常赞同GreenUP项目，作为美国密歇根州立大学城市社会学的教授并著有《建筑全球化——中国城市中跨国建筑的生产现状》一书， 他在此书中指出"在书中所提到的多功能绿化种植项目在南半球的一些特大城市都已起到了广泛的作用，所以中国及印度作为世界两大发展中国家，尤其应该实行城市绿化项目。"

除去现实原因，中文中的"树"字（图2）与由达芬奇设计的GreenUP 的树型标志（图3）之间的相似性无疑也在呼唤着中国应该开展GreenUP此类项目。

合作资质及项目开发

GreenUP项目旨在提升落后的设计方法（图4），对改造地点重新提出具体的规划。实行这个项目的规划，首先需要的是要拥有一个怀有高标准理念的投资者，接着是要选择改造地点，然后是对开展这一项目达成一致意向。

当这些前提条件得到满足后，请联系CrossingLab.com的负责人，Giacomo Pirazzoli 教授，电子邮箱为crossing@GPspace.org。

关于GreenUP — A smart city书的推荐

"我们需要的不仅仅是更好的进行经济建模和对社会经济数据进行更严格的计量经济分析，我们更需要的是有远见的将城市发展的未来前景与历史结合起来。"——Amitabh Kundu博士，经济学教授，印度新德里贾瓦哈拉尔·尼赫鲁大学

"非常高兴能读到这本书：我们多次提出了那些无法操作的方法，或者为实现可持续发展而列下抽象的目标。事实上，解决人口及建筑密集需要适当的方法，但是有太多象征性的呼声使这项工作原地踏步。相反，在这本书中，我们看到了想法、远见，我们更要从实际的角度来解决问题。城市绿化提升以及大都市的发展需要可行性高且能实现可持续性发展的项目。"——Tommaso Vitale教授，巴黎政治学院主任，主要教授大都市管理课程

"提升绿化项目主要是在陈旧的建筑物及场所中运用其原有的材料进行整修，这也意味着这个项目需要当地社区的参与及实行低成本的干预措施。其中最主要的方法是如何将现有的建筑及材料全部进行再利用。垂直的景观植被不是用来促进新的发展，而是为了重现目前的状况。"——Giulio Giovannoni博士，佛罗伦萨大学城市规划教学小组成员

"住宅环境，与任何的日常环境一样，都逐项地不断地在进行着不可阻挡的改变。其变化的模式也反映了社会群体——一方面是家庭，另一方面则是社会及经济主体。由于我们很好的认识到了变化的模式及隐藏在其背后的那些约定俗成的规则，所以我们提出的创意及实行的干预措施对环境协调及可持续发展做出了不可磨灭的贡献。"——Stephen Kendall教授，开放性建筑的创始协调人

"超过一半的世界人口都居住在城市中心,并且气候变化也加剧了城市食品安全危机,这也是现阶段面临的新挑战。"——Cecilia Tacoli博士,英国气候及发展国际研究中心

"GreenUP — A smart city这项具有远见的项目证明了当人们将城市所面临的挑战视作可以互通共享的机会时,也为未来的城市发展带来了更多的可能性。"——Majora Carter,城市复兴策略专家,布朗克斯,纽约

"GreenUP - A Smart City" aka Preservation to be Conceptualised

Giacomo Piraz Pirazzoli

"GreenUP - A Smart City" is a cross-media research coordinated by prof. Giacomo Pirazzoli (iCad_International Course on Architectural Design – University of Florence, Italy) actually produced by CrossingLab.com think-tank.
A highly flexible and affordable vertical landscape network, GreenUP focuses on Cultural and Monumental Heritage as expanded fields.

GreenUP adaptive framework allows working either in metropolitan contexts for sneaking in to connect existing green fragments, to deal with new planning process to start from scratch either.

A truly alternative vision to Global Branding then to star-architects dictatorship, GreenUP site-specificity is to become a pivotal way for sustainable settlements to harmoniously merge contemporary architectural design with preserved traces from the past.

AIM
A "think global/act local" bottom-up tool, GreenUP aims at enhancing environmental qualities such as livability, then at decreasing air pollution while increasing Co2 balance etc.

GreenUP may also provide food security after fruits and vegetables to grow first, then to socially share within the community.

Key-Words
GreenUP keywords include: Carbon Footprint Reduction, Common Goods, Urban Farming, Recycling, Energy, Biodiversity.

Research Background
GreenUP is a holistic work started by about one hundred researchers – students, architects, botanists, engineers, agriculture experts, urban planners etc. – all working across disciplines for the past five years in a volunteering capacity.

Under "GreenUP - a Smart City" conceptual framework several case-studies have been processed worldwide jointly with partners institutions in Europe, as well as looking towards Developing Countries.

A comprehensive research preview has been presented first at Governing the Metropolis International Conference, Paris 2012.

After the fully successful "Urban Connectors" TEDxConference (IMAGE 1):

http://www.tedxhamburg.de/greenup-a-smart-city-giacomo-pirazzoli-at-tedxhamburg-urban-connectors, GreenUP seminars and presentations took place across the US, then at AAPI_Australian Asian Pacific Institute (Curtin University, Perth AUS) as well as at EU/Brazil Climate Change Conference, Rio de Janeiro 2015 etc.

At present the follow-up for the research to be fully applied includes several international start-up partnerships.

Cross-Media Bundle

The whole GreenUP cross-media project includes printed book (Allemandi International, Turin-London-New York 2013) and augmented ebook with links.

A side-video by movie director Filippo Macelloni, with original drawings by artist Agnese Matteini has been produced: https://vimeo.com/87993255

The website www.CrossingLab.com offers a significant preview.

Greenup In China

When endorsing GreenUP, China-born Dr.Xuefei Ren, a Professor of Urban Sociology at the Michigan State University (USA) and the author of Building Globalization - Transnational Architecture Production in Urban China stated "The multi-purpose green settlement projects presented in this book have wide implications as common goods for megacities in the global South, especially India and China."

Beyond this real and well reflected issue, the similarity between Mandarin-Chinese ideogram which stands for "Tree" (IMAGE 2) and Leonardo Da Vinci's designed GreenUP tree-logo (IMAGE 3) undoubtedly calls for a GreenUP release to be developed in China.

Partnership Candidature / Project Development

In order to conveniently upgrade an early Network, Hardware, Software strategy (IMAGE 4) up to a GreenUP site-specific design – an investor with high standard ideas is needed first, plus a site, plus a program to agree on.

Whenever such pre-requisites are satisfied, please contact CrossingLab.com director, prof.Giacomo Pirazzoli crossing@GPspace.org

Some "Greenup - A Smart City" Book Commendations Worldwide

"We need not just better modelling and more rigorous econometric analysis of the socio-economic data but also visionaries who can link up history with future perspectives of city development."   — Dr. Amitabh Kundu, Professor of Economics, Jawaharlal Nehru University, New Delhi

"What a pleasure to read this book: so many times we have just accounts of what is not working, or lists of aims and abstract goals of sustainability. Actually population density and building concentration require appropriate solutions, but too many actors voice standing still. On the contrary here we have ideas, visions and real perspectives. Feasible, sustainable projects are available to GreenUP urban and metropolitan developments."   — Prof. Tommaso Vitale, Director, Master Course Governing the Metropolis, Sciences Po, Paris

"GreenUP mainly elicits as its row materials to be worked on decaying buildings and situations, implying low-cost interventions to be implemented with the participation of local communities. One of its main principles is the complete reuse of existing buildings and materials. Vertical vegetation is not used to promote new developments but to re-signify existing situations."   — Dr. Giulio Giovannoni PhD, Urban Planning Teaching Unit, University of Florence

"Housing environments – like all everyday environment – are constantly, inexorably transforming, part-by-part. The patterns of change reflect social groupings – families on the one hand, and social and economic bodies on the other. Conventions rule. We do well to recognize these patterns of change and the conventions underlying them, so our inventions and interventions contribute to environmental coherence and sustainability."   — Prof. Stephen Kendall, Founding Coordinator Open Building

"More than half the world's population lives in urban centres, and urban food insecurity is an emerging challenge that is exacerbated by climate change"   — Dr.Cecilia Tacoli, International Institute for Environment and Development, UK

"The visionary work in "GreenUP - A Smart City" shows what can happen when one views the challenges of urban areas as opportunities to share."   — Majora Carter, Urban Revitalization Strategist, The Bronx, NY

项亚量
Xiang Yaliang

受教育情况
1999-2001年　于浙江大学建筑系城市规划专业攻读硕士学位，后因出国留学中断。
2001-2002年　于柏林工业大学城市与区域规划专业攻读硕士学位，后转专业。
2002-2006年　攻读柏林工业大学建筑设计专业，获得硕士学位。

工作经历
2003-2006年　工作于德国本哈特·温克教授建筑事务所
2006年—至今　德国本哈特·温克教授建筑事务所驻中国代表
2007年—至今　中国美术学院中德学院担任建筑专业讲师
2015年—至今　中国美术学院中德学院与安哈尔特应用技术大学合作"建筑遗产保护硕士项目"负责人

Study
1999 - 2001　Master Course of Architecture at the Zhejiang University, Department of Architecture and Civil Engineering
2001 - 2002　Master Course of Urban Design at the TU Berlin
2002 - 2006　Master Course of Architecture at the TU Berlin, graduation as Dipl.-Ing. (graduate engineer)

Professional career
2003-2006　Coworker at the Prof. Bernhard Winking Architects BDA, Berlin und Hamburg
Since 2006　Representative of Prof. Bernhard Winking Architects BDA in China
Since 2007　Architectural Docent at the Chinesisch-Deutsche Kunstakademie of China Academy of Art
Since 2015　Course coordinator for the international study course in Monumental Heritage at the Chinesisch-Deutsche Kunstakademie of China Academy of Art, in Cooperation with Anhalt University of Applied Sciences in Dessau

新区新建筑与历史文脉
——以杭州双城国际项目为例

项亚量

【摘要】分析杭州双城国际项目设计过程,特别通过对比国内和国外两个方案,让人们看到一个令人尴尬的现实,即:某国内事务所不重视本土文化,盲目追求西方时尚,而某国外事务所尊重并准确把握项目所在地的文脉,并用高超的现代手法加以实现。通过介绍国外事务所在双城国际项目具体的概念,尝试剖析席卷全球的现代建筑在精神层面上也能和中国传统建筑文化合拍和共鸣。

[Abstract] After analyzing the design process of Hangzhou Twin Cities International Project, especially comparing the domestic scheme to the foreign scheme, people see an embarrassing fact: a domestic office didn't attach importance to local culture and blindly pursuit western fashion, but another foreign office respected and correctly grasped the culture in the place where the project is located and realized its purpose with modern techniques. To introduce the concrete concept of Twin Cities International Project of a foreign office and try to analyze the modern architectures having swept the globe also can be in harmony and produce resonance with Chinese traditional architecture culture at spiritual level.

中国急速的城市化举世瞩目。诺贝尔奖获得者Stiglists曾把"中国的城市化"和"美国的高科技"称为21世纪人类发展的两大进程。城市化意味着大量的人口从乡村到城市的转移。另外一方面城市也必将快速扩张,并占用大量城市周边的非建设用地和乡村用地。

过快的城市发展的弊端和后果除了反映在经济、生态和社会领域,更是自"文革"以来对城市的文化脉络的新一轮强烈的冲击和破坏。"这个世界城市史上绝无仅有的全国性的'造城运动',已经将我们的大大小小的城市全部卷土重来一次,抹去历史记忆,彼此克隆,最终像蚂蚁一样彼此相像。"[1]

当前建筑师所面临的大量的设计项目是在一片刚刚从城郊农用地拔地而起的城市新区内的项目。这些项目所在场地由于几乎没有直接可以参照的历史文脉的存在,往往特别容易受到商业化的冲击。在商业化面前,所有建筑都是商品。因此无数靠出奇制胜,甚至没有底线的花哨的外形、新奇的风格一时间充斥新区。人们对此往往认为新区反正是没有历史记忆和文脉的文化荒漠,因此也就对此熟视无睹。

然而吴良镛先生在《1999年国际建协第20届世界建筑师大会"北京宪章"(草案)》中对现代建筑和地区文脉之间的关系做了深刻的阐释,并且提出"现代建筑的地区化,乡土建筑的现代化,殊途同归,共同推动世界和地区的进步与丰富多彩。"[2]那么在城市新区的现代建筑如何做到地区化呢?

本文要介绍的双城国际项目位于杭州市滨江区。其地块南到滨盛路,东为江汉路,北至秋水路,西靠通江路。北侧连接绿地,面向钱塘江。用地范围地势平坦,市政设施完善,周围道路基本形成,交通方便,自然环境优美,是杭州跨江发展的重点开发地区,也是滨江区的行政中心区块。项目用地面积约3万平方米,建筑面积约12万平方米,高度100米,主要功能为办公楼。地块环境具有典型的以汽车发展、交通设施为导向发展的城市新区空间的特征,其城市尺度巨大。在设计之初,场地周围除了建设完成的道路外都是大片农田。

建筑师首先从大尺度的"时间—空间—人间"(吴良镛,1999)的角度对杭州城市文脉进行了分析,力图在滨江这片新城上找到与杭州城市特征相呼应的历史和文化关联。在总体布局上,建筑师提取了中国传统院落建筑的精髓。中国最典型的院落结构将建筑与城市之间构成了一个强烈和独特的空间序列—开放空间-半开放、半私密空间-私密(内部)空间。在这围合空间中体现了"空的主导,春夏秋冬的轮回,日出

日落的更替，阴阳运转，五行生克。这时间性特征，充满动态性、音乐性乃至空无的悲壮性……西方文化的空间与时间是分开的，而在中国文化中，时空合一。"[3]

在对杭州城市特征的总体把握上，建筑师认为，杭州与中国其他许多城市相反，不是以众多的高楼大厦组成。其真正的城市特征正是通过历史性的建筑遗存和中等高度的块状结构所勾画的。所以建筑设计中，建筑师试图将不同的尺度联系起来，使之紧密连接，从而得出将双城国际项目设计成为具有雕塑感外形的创作思路。建筑物由两部分建筑单体构成，犹如中国传统文化中的阴和阳。它们围和着一个向公众开放的水上广场。两幢八至二十六层的大楼就像斯芬克司狮身人面像一样威严地耸立在钱塘江畔。台阶式的造型使底下的楼层拥有大面积的屋顶花园。

通过透视上的重叠，两幢建筑从不同的观察角度，会变化出截然不同的空间形态：从钱塘江正面，或从生态公园和滨盛路观察，它们是开敞和通透的；而在钱塘江上从侧面（比如在钱江三桥和四桥）观察，却是封闭和巨大的。这种不同的表象变幻将使建筑物在钱江新城中显得极具个性。

同时，这样的围合布局有利于将钱塘江的江景延伸到城市中来，而不是为了自身"最大化"一线江景而做成屏风状，从而割裂城市和自然之间的联系。顺便值得一提的是，在选择最终方案前，甲方曾经花了一年半的时间讨论两个方案的取舍。其中落选的方案，见右图。

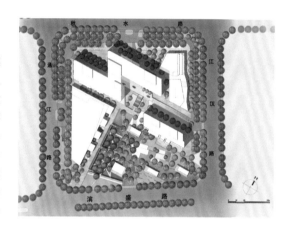

从这个方案总体布局上，可以体会出这个方案的建筑师自觉不自觉地反映出其受以自我以及个人为中心的西方文化的影响。"（在西方文化中）设立（标志）成为无限空间中生存环境的标志，聚落的标志，城市的标志……在西方文化中，标志成为一种归宿，成为上帝的代名词，膜拜的对象，成为凝视的目标。在凝视的历史中，时间固化、目标锁定、透视学产生。凝视因而成为西方人观察世界的根本方法……这凝视充满了静态的特质、刹那间的时刻性……"[4]因此从根本文化气质上，这个方案更加代表的是强调自然与人对立的西方文化，而与中国传统"天人合一"文化观念相悖。这点在他的设计草图中得到充分体现。见下图。

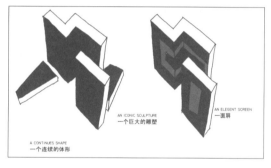

另外，由于过长的体型，导致内部平层交通空间面积过大，得房率降低；再加上折线形的设计对于高层建筑的结构来说也是较大的挑战，造价因此飙升。最后此方案落选是理所应当的。

有意思的是，在立面造型和材料上，两个的方案也形成了强烈的对比。落选方案外立面采用了全玻璃幕墙结构。设计师大费心思将甲方高科技企业特征（0和1的二进制数码）和中国传统装饰图案结合在一起，设计了这种颇具时尚感的外墙纹理。但这只是一张简单挂在结构外面的一层肤浅表皮，并没有和建筑本身发生任何的联系。而且这种全玻璃幕墙的设计已经被证明是非常不利于环保节能，另外还会产生光污染，影响周围的城市环境。

而入围方案抛弃一切浮华的装饰，忠实地反映建筑框架结构本身。恰恰是这一点及其现代主义的做法又和中国传统建筑合拍并产生共鸣。因为"现代建筑虽然蜕变自西方的古典建筑，但它是摆脱西方传统建筑的束缚而发展起来的，比较起来似乎和中国古典建筑在原则上就更为接近。'框架结构'就是其中的一个最主要的共同点，一切建筑构图的问题都是由此展开。"[5]

在立面的细节处理上本方案也非常有特色。其外表有意地将杭州传统的建筑方式运用与现代的形式中。建筑高度的划分通过外立面墙体的回退（即柱子下大上小，而玻璃下小上大），产生雕塑性的效果。表现了

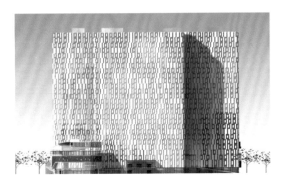
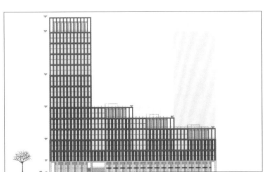

结构受力上、受地心引力影响上小下大的基本规律。同时这样的造型也加强了建筑竖向的透视效果。

外立面除了玻璃外，设计师采用了中国传统建筑常用的红砖。但是砖墙不再是由人工堆砌而成，而是采用现代化的混凝土预制工艺。红砖外墙是具有生命力的，而且"愈老弥坚"。特别是在风雨过后，红砖表面吸收水分颜色变深。然后随着阳光的照射，受阳面的砖首先变浅，然后逐渐扩散到整个外立面。红砖外立面的使用还在满足室内足够光照的前提下，避免了室内过度受阳光照射，从而降低了空调的能耗，并且避免了光污染对城市周边的影响。然而在项目实施的过程中，受到当时技术特别是预制混凝土技术，以及造价的限制，项目外立面最终采用了灰色花岗岩。

上面提到的花岗岩是国内市场上最常见的芝麻灰。就是在对着再普通不过的石材的应用上，也体现了设计师独到的匠心。设计师将同样石材表面的抛光程度分为三个等级。抛光等级最高的用在外墙墙面，中等抛光的用在室内地面，毛面的花岗岩用于室外景观地面铺装。这样将建筑、室内、景观有机地联系在一起。

最后我想介绍的是这家对中国传统空间精神有着如此深透的理解，又能用现代建筑的语汇表现出来的设计单位，并非来自国内。它是来自德国的本哈特·温克教授建筑事务所。本哈特·温克教授是来自汉堡美术大学建筑系的教授，现年已经八十高龄，还仍然奔波于世界各地。他一生著作颇丰，获得过众多奖项。笔者有幸在双城国际项目与他合作。他每每谈到自己的设计哲学时就显得非常郑重。他总是强调自己是"现代主义者"，他对建筑的理解是"建筑永远是人类历史、城市肌理和物理空间的组合"，而他的每一次设计的出发点都是"找出设计项目的本质、实现业主的愿景、并且回顾过去，从中得出结论—未来将会怎样。"双城国际项目前后持续了5年时间，期间无论是甲方还是国内施工图配合单位都深深为温克教授的人格魅力和执着精神所感动。

通过这个项目，我深刻体会到西方建筑自上世

纪初由装饰主义（art deco）和现代主义两条主干发展至今的不断演变。两者都试图抛弃传统，而前者不忘从传统以及外来文化中提取灵感，后者则提倡与传统革命性的决裂。两者之间又能够相互借鉴，从而生化出新的建筑形态。人们是否能从双城国际项目立面形态上隐约体会到希腊帕特农神庙的精神呢？

最初西方建筑通过吸收非洲、美洲、亚洲的符号来丰富和发展自身的建筑形式。经过一个世纪后，已经从引进表面的、新奇的、装饰性的符号，逐渐深入到对各种文化的核心，并融会贯通。恰如林语堂所说的"两脚踏东西文化，一心评宇宙文章"的精神。这对于中国建筑设计本土化有着深刻的借鉴意义。

注释

[1]《灵魂不能下跪：冯骥才文化遗产思想学术论集》，冯骥才 著，宁夏人民出版社，2007，第12页
[2]《世纪之交的凝思：建筑学的未来》，吴良镛 著，清华大学出版社，1999，第9页
[3]《园院宅释：关于传统文化与现代建筑的可能》，李劲松 著，百花文艺出版社，2005，第44页
[4]《园院宅释：关于传统文化与现代建筑的可能》，李劲松 著，百花文艺出版社，2005，第41页
[5]《华夏意匠——中国古典建筑设计原理分析》，李允鉌 著，天津大学出版社，2005，第27页

黄晓菲
Huang Xiaofei

受教育情况
2000—2004年　中国美术学院环境艺术系，获学士学位
2004—2007年　保送至中国美术学院环境艺术系，获硕士学位

工作经历
2007至今　工作于中国美术学院，现为中国美术学院中德学院专业教师，讲师。从事建筑遗产保护方向的教学、研究和实践。

Study
2000-2004　Study Environmental Design at China Academy of Art Bachelor's degree
2004-2007　Study Environmental Design at China Academy of Art Master's degree
　　　　　　Academically rewards
First Prize Scholarship, China Academy of Art
Zhejiang Outstanding Graduates, China

Working experience
Since 2007　Lecturer at the Chinesisch-Deutsche Kunstakademie of China Academy of Art. Engaged in the Monumental Heritage protection of teaching, research and practice.

凸显城市身份定义的景观实践教学之道
——以杭州工业遗存景观更新设计为例

黄晓菲

【摘要】 毕业设计在高等职业院校艺术类设计教学环节中扮演着收纳精髓、启迪创新、展现睿智、承上启下的作用。景观建筑学毕业设计课程从选题确定到模式设计、方法选择既要考虑技术手段的充分展示，更要注意人文精神的全程浸润。以建筑遗产保护为专题导向的毕业设计课程以"场地认知和意识再现"作为专业教学核心，让学生在特定场地中寻找地缘特性，并以此作为一系列设计行为展开的身份特征和地理原点。通过教学思路的变革与教学方式的更新，引导学生尊重本土、正视本土、发掘本土，调动各种设计语言和技术手段再现学生对景观的见解和思考，建构学生对景观建筑物理效能、社会效能、美学效能、历史效能的认知模式。

[Abstract] Graduation design plays the role of gathering essence, enlightening innovation, showing wisdom and connecting the preceding and the following in the teaching of art design in higher vocational colleges. For the graduation design course of landscape architecture, not only the sufficient showing of technological means for topic selection, model design method selection shall be considered, but also the infiltration of humanistic spirit during whole process shall be paid more attention. The graduation design course with architectural heritage protection as thematic orientation takes the "site cognition and consciousness representation" as the core of professional teaching to make students seek geographical characters in a specific site to be the identity characteristics and geographical origin of a series of design behaviors. Through the reform of teaching thought and the update of teaching methods, we guide students to respect local culture, face up to local culture and explore local culture, to mobilize all kinds of design languages and technical means to represent the insights and thinking of students on landscape and construct students' cognitive pattern of physical efficiency, social efficiency, aesthetics efficiency and historic efficiency on landscape architecture.

一、引言

毕业设计是高等职业院校艺术类专业人才培养教学过程中的最后阶段，教学时间数倍于其他单元专业课程，在整个大学教学环节中占有较大比重，约占学生专业教学总课时的20%。这一课程的设立，既是高等职业教育理念与教学效能的综合检验，也是学生学习能力与创造能力的全面表达。高等职业教育今天培养的应用型人才素养直接连接明天的行业风貌，景观建筑学毕业设计的目标定位直接影响未来国内景观设计从业者的设计意识与设计水平。

尽管毕业设计举足轻重，但也遭遇了多重矛盾。例如授课教师多与一的矛盾。高校艺术类设计专业的分段教学体系要求此前的各个专业科目由多位个性不同的教师分担，而毕业设计课程作为一门综合且独立的教学科目则只能由一个教师负责。该教师不一定是前面分段教学的老师，故对毕业设计教学对象之前的教学内容与接受特征不甚熟悉，无法形成对教学质量的持续关注与把握。又如在教学实绩展示形式与内容的矛盾。方家皆知，艺术类职业院校景观设计学专业毕业设计是学生与老师共同经历的一场智力与体力的大比拼。在教学时间与投入资金极其有限的情况下，教学重点容易流于对具有视觉冲击力的设计展示效果的追逐，而非耗时耗力地探索设计本源。再如教学内容方面零散与整合的矛盾。学生在单元专业课程中积累的大多是应用型技能和相对零碎的知识点，而毕业设计课程所要求学生具备的则是化零为整、由表及里透过现象看问题本质的设计思辨能力。因此，毕业设计课程的教学定位、教学模式以及教学方式成为亟待讨论与思考的重要话题，需及时做出合乎教育规律和本质的理性回答，否则学生就会在经历漫长的设计周期之后身心疲惫，依然无法清晰明确回答原本属于景观设计的基本问题。"何为景观设计？""景观设计为何？""如何景观设计？"面对这些理应侃侃而谈的诘问，他们只有困惑、混沌、疑虑。带着这些茫然步入职场，能够给社会奉献什么景观精品自然不言而喻。

　　坚持以技术应用能力培养为主线的培养模式，是我国艺术类职业院校的共识。高职景观设计毕业设计课程的教学模式一般开始于设计流程的介绍，结束于效果图的制作表现。整个教学环节基本以教师为中心进行授课，学生通过大量制图训练不断巩固软件的应用操作能力，老师讲解改图。但从艺术教育、人才教育以及景观设计的内在规律出发，职业教育无法跳过对学生景观专业根本认知的培养，补足其缺乏设计理念思辨性和逻辑思维的缺憾。为此我们需要打造一门强调符合景观建筑专业特点，将课程教学核心知识点从应用技能教授转为过程推导研究，以进入真实场地为基础，以培养设计思辨思维为核心的专业核心课程。学生只有掌握了这种本质的设计方法论和设计思维后，才能在以后长期职业生涯中目光敏锐、思路开阔、技艺娴熟、后劲充足。

二、定位

　　为回应现状、适应未来，笔者自2010年开始，连续6年先后在4个教学班级毕业设计教学实践中展开"以杭州工业遗存景观更新设计为例"的教学改革研究。聚焦杭州本土城市发展中的建筑有机更新问题，结合受到国内外高度关注的"建筑遗产保护"话题，引入"杭州工业遗存"这一专题性主题展开景观建筑学毕业设计教学改革研究，从毕业设计课程的选题、教学思路、教学模式等方面进行实验性探索和教学尝试，以期扩展学生对于建筑遗存的认知边界，从根本上引导学生掌握理性思辨的设计方法。

　　该课程是职业院校景观建筑学的核心课程（foundation Studio）。改革的前提是对两个问题的明晰：什么样的设计选题能够兼顾"真实"同时能够引发对于景观建筑学的深层思考；这个阶段的学生已经具备什么专业基础，他们缺乏的核心知识点是什么。对这两个问题思辨的结果势必导引出"怎么教"。一切教育思路、教学方法的改变都必须建立在坚实的教育学基本规律基础之上。强调动脑与动手相统一的实践教学是高等艺术教学的特色与必须，从单纯知识传授向注重能力培养转化亦是现代高等教育发展的大趋势。只有这样才能实现"像哲人一样思考，像匠人一样劳作"。我们培养的学生设计能力与创新能力，将集中体现在面对具体问题的分析与处理之上。改革后的课程教学核心是一种基于进入实际场所的工作方式，由真实场所引发学生对于场所特有的地缘要素、历史要素、文化要素等综合思考，运用经由严谨、复杂、科学的设计实践推导训练，培养学生逐步建构科学理性与人文理性相统一的专业评价标准。围绕这一基本理念而进行的课程改革，将能力培养与素质教育置于优先地位，既符合艺术教育的规律，也顺应高等教育的趋势，是"合规律性"与"合目的性"的统一。

　　景观建筑学作为一门服务于社会的应用型综合学科，其教学特征与专业特点应当是一致的：一方面，当下的社会热点问题应立即成为教学难点或讨论要点。学生可以因对社会问题的高度关注而引发设计思考、提出设计策略，在社会责任感、参与感与使命感的驱动下由景观设计之道寻找城市身份定义与城市文化源头；另一方面，景观建筑设计有其十分明确且严谨的内在规律。无边的遐想与语汇的偏好都必须接受规律的筛选，只有"合规律"才有生命力。只有将理性精神、批判思维贯彻设计全过程，才能得出科学合理、构思奇妙的设计策略。"直面现实"与"依循规律"要求毕业设计课程无论在设计选题还是教学定位方面都力求凸显"真实"。而要把握这深邃的"真实",学生则首先得知晓"场地真实"以及由其所承载的"文化真实"。

选择杭州工业遗存景观更新设计专题原因有四:其一,杭州作为中国的六大古都之一,早在南宋时期就是制造业中心,"造船业、丝织业、瓷器与纸张的制造在南宋尤其突飞猛进。"(《中国大历史》,黄仁宇著,三联书店,1997,第163页。)近代之后更是浙江最重要的经济、文化重镇。随着近年经济水平的提升、产业结构的转换,旧有企业关、停、并、转已成常态,因此在城区留下了一批数量可观、保存良好、值得关注的工业建筑遗存。这些工业建筑遗存成为研究城市有机更新问题和重寻城市文化特征的适合切入点。其二,每个大学的任何核心课程都需要经历多年的逐步优化和深化构成院系教学体系的学术重要骨架。本课程也遵循相同规律,在杭州工业遗存的大课题下通过多届学生对一个设计专题的追踪研究,获得了许多可以环比的数据,并借此调整、发展、优化成逐渐固化的课程教学模式。其三,整合自然资源并协调人与自然的关系是景观建筑学的核心所在,进入真实的自然事物是景观建筑学的必然教学方式。面向真实历史遗存,基于真实场地工作,以真实的场地为课堂,探讨地形(Landform)、植被(Planting)、水(Water)、建筑(Architecture)这些关键景观元素的地位与作用,可以使得教学生机勃勃充满活力。其四,本课程设置的对象为美院毕业班学生,他们的技能优势与思维特长是绘画与表现,但关于建筑哲思层面的思考和研究方法论的掌握则是普遍的短板与缺陷。建筑遗产保护课题恰好提供了一个非常好的契机,使我们在教学上能够有针对性地实现"补阙拾遗",让学生在设计实践中了解思辨能力的重要、建筑社会属性的重要、建筑社会学思考的重要。

基于这样的课程定位,笔者要求学生必须经历"走进去""走出来"两个过程。"走进去"是带领学生走入图书馆和档案馆,通过阅读和查询史料扩展其社会学、历史学、政治学等人文通识学科的知识边界。景观建筑学(Landscape Architecture)是一门为人类生存环境服务的综合性设计学科,城市发展过程中的每一次重大社会、经济转型都会催生景观建筑学的新思考。现代城市在迅猛发展中之所以经常出现"千城一面""千村一面"的现象,城市文化特征缺失与城市身份定位缺失是普遍原因。工业遗址、废弃工业区改造和棕地景观再生之所以成为当前国内外同仁讨论的热点,源于对工业建筑遗产在建筑遗产中地位的确证。对工业建筑遗产的保护与更新,既是对建筑遗产保护认知的深化,也是对城市历史的追溯。只有在这样的高度建构对此项工作意义的思考,才能使学生形成从事该项目研究与实践的价值信念与理论自觉。

"走出来"是带领学生走出教室,进入真实场地的探索,实地测量、实地考察、实地教学是景观建筑学最基本的教学方法。让学生直接置身实地,可以激发学生直面现实的活力、考察学生运用书面知识解决实际问题的能力、调动学生各种感官感知具体场景的热情。学生通过对真实基地的走、看、记录(快写、慢写、摄影、拼贴),可以变晦涩的感念为鲜活的体验,使得课堂中的基础训练在观测与记录中活化。

三、意义

鼓励多样化地理解文化遗产的概念和评价文化遗产价值的重要性,已经成为一种新的国际学术潮流,并对国内学界产生了积极影响。景观设计教育者的重要使命是将这些有益的思路引入课堂,教会学生透过问题看本质的溯源能力,采用各种训练潜移默化引领学生快速把握本质能力的提升。任何建筑的出现与变迁背后都有着非常复杂的社会学与历史学原因,对任何建筑的解读都需要借助多学科的支持。通过项目实践教学形成建筑社会学、历史学观念,社会学、历史学、地理学、哲学、甚至建筑师个人因素都与建筑互为因果,最终达到在设计教学过程中将教授简单的设计手法与表现技能上升为训练建筑哲思思维的目的。

　　随着经济全球化步伐的加快，产业结构发生世界范围的调整。东西方各国相继进入后工业化时期，超越机器工业大生产时代已经成为思维与实践的必须。旧时的工业建筑被纳入"历史遗迹"这一概念之中，和其他所有名胜古迹联系在一起，成为一种新型文化遗产，即工业遗产。2003年7月，国际工业遗产保存委员会就在《关于工业遗产的下塔吉尔宪章》中对工业遗产进行了明确定义。(《西方废弃工业区的更新改造对策和实践》，马航 等著，科学出版社，第12页。)自上世纪60年代欧美发达国家就已经先于中国结束了工业革命进程，进入后工业时代，开始了城市老工业区更新改造的研究，在工业遗产再利用方面，起步大大早于我国。其中英国、德国在这方面的理论研究较为系统和成熟，相应的设计实践经验也更为丰富。德国安哈特应用技术大学建筑系多年前开设的"建筑遗产保护"专业方向在德国专业排行榜上名列前茅，对建筑遗产的开发策略和保护方法进行深入研究，获得了较为客观全面的理论体系。与此同时，作为一种有可持续性的发展方式，"工业遗产旅游"被视为城市产业转型的触媒被广泛关注。大量的后工业景观案例，例如多特蒙德园林展、慕尼黑园博会矿区公园、科特布斯矿区恢复、纽约高线公园以及德国 Zollverein 工业艺术综合体等众多旧工业遗址改造项目更为国外后工业景观的研究提供了丰厚的基础。工业建筑遗产"从艺术历史或科学的角度而言，具有显著的广泛价值。"(《传统之寓言》，Francoise Choay 著。)

　　与国外相比，在后工业景观方面的研究我国尚处于起步探索阶段。面对各城市中的工业建筑遗产，虽然在设计理论界与实践项目中已经有了一些成果，但将其纳入高校景观设计学教学研究领域尚不多见。选择杭州工业遗存景观更新设计主题展开的毕业设计教学研究，具有多重现实意义。其特色和重点是将杭州本土的工业建筑遗产这一真实自然的场地视为一座将设计实践课程过渡为理性思考与实践结合的综合课程的桥梁。

四、重点与难点

　　在设计这一课程时，有许多教师更关注最终效果的展现效果，故在时间分配与精力分配上偏重展示制作环节，而将真正的设计一笔带过。这样的教学设计并没有体现毕业设计课的特殊性与综合性。改革后的毕业设计课程力图改变以往教学中"重结果、轻过程、无调研"的弊病，以"场地"为核心的户外课堂教学，让学生面对一个真实的、能反映历史演变与社会经济发展的工业历史遗存场地。在教学时间安排上，调研分析过程与设计环节之间的时间配比为2:1，保证学生在设计前期对于场地特性、使用对象需求、对象行为等进行充分分析。重点关注景观建筑学所涉及的核心知识点，即设计前期的哲思部分和理性研究。毕业设计的授课时间一般贯穿最后一个学期，为期12周。每周课时数为18学时，共计216学时。为充分利用有限课时，笔者采用让学生提前介入的工作模式。在上一学期结束前就布置课题、开列参考书目，让学生放假前进行第一次场地考察与同类型项目的实地调研、在假期中阅读有关文献，使学生有充裕的时间对项目性质和项目条件的熟悉认知。开学后即用7个课时完成理论教授，对场所的考察分析与具体设计方法的讨论。与传统的毕业设计更多地借助计算机辅助技术解决表现手段不同，新的毕业设计课程从设计之初就鼓励学生运用手绘、摄影、拼贴等综合方式从场地中提取设计元素，表达、纪录场地，获取景观的延伸表现，模型制作成为贯穿始终的设计手段。

　　余下的200多学时教学计划是以真实的场地——杭州本土工业历史遗产为核心展开。6年间笔者相继选择了杭州重型工业机械厂建筑遗产、杭州双流水泥厂建筑遗存、杭州钢铁厂薄板分厂等工业遗存作为系列专题。全班同学以4－5人为一组展开工作。根据认知的过程，分为项目调研（2周，毕业设计之前进行，个人工作）、资料搜集（2周，个人工作）、场地熟悉（3周，个人＋团队）、场地探索（4周，个人＋团队）、设计批评（3周，个人＋团队）共五个阶段。

每个阶段都有明确的教学要求与任务安排，尤其要求学生多次、不断地回访场地，以期在设计的不同阶段获得不同的场地感受，不断反思设计。这其中包含如下几个关键：

1.鼓励学生做一手研究，开展以杭州政治、经济、历史、社会为导向的建筑背景研究。

工业建筑遗存隶属于建筑遗产范畴，因此我们引入"建筑遗产保护学"的专业方法对工业建筑遗存进行先期调研。将学生拉入理性分析的思维路径，目的在于让学生意识到追溯表象背后内在原因的重要性。只有揭示出建筑背后的生成规律，才能有的放矢大胆取舍。先期准备工作包含两方面的内容：从建筑遗产保护角度来看，了解建筑遗存的历史成因尤为重要。试图由一系列的发问拼凑出建筑的原始面貌。例如：建筑在什么样的历史条件下建成、当时的经济条件是怎样的、当时是否有重大的社会事件发生等等。另一方面则是相对具象清晰的建筑本体分析。例如对建筑遗产的现存状态进行测绘调研，完善和创造性组织这些资料图纸。要求每位同学对现存的工业建筑遗产从整体到局部进行照片记录、编号，创建照片档案。然后运用色彩分析手段分析、绘图，在建筑立面上标注不同建造时期，为后面景观建筑设计的介入做准备。

对于任何建筑遗产项目都要求学生必须完成重要的准备工作研究（Important preliminary studies），工作内容主要有三个方面：1.建筑研究（Building Research）；2.历史文本搜集（Inventory plans/ Architectural study/ Documentation）；3.场地地图分析（State mapping）。作为项目的基本调研（Basic research），要求学生从档案馆、报纸、相关办公机构获得与项目有关的地图、照片、建筑档案、原始平面图纸、城市历史等第一手原始资料。这些一手资料能够给学生带来一些原创的发现，用这些新材料去了解当时的时代背景。

2.基于真实场地的四次场地工作。

整个毕业设计过程中要求学生至少完成四次场地工作：第一次场地工作是对场地初识，包括对于场地中雨水、植物、建筑、光线、风向等最直观的感受，完成场地分析与调研，形成场地报告；第二次场地工作重在发掘场地建构问题，包括建筑之间的空间关系、建筑结构特质，特殊的地形和尺寸，显著的历史层次和悠久的时间变迁等，同时考虑包括光线、纹理、风、水、植被、声音等在内的自然要素。继而根据场地地缘文化进行整理分析，总结归纳项目特质，确定设计主题，进行场地分析汇报；第三次场地工作伴随设计工作同时进行；第四次场地工作设置在整个设计环节的中后期，带着设计重新回到场地探讨空间。

在无数次场地穿越过程中，学生身兼调查员、设计者和观察者的身份，每一次的场地调查都能唤起他们对场地的不同层面的审美判断，经由个人体验上的转换之后的场地再现方能成为一种深思熟虑的设计手法。在这个过程中，我们要求每位同学完成记录场地的表达，主要手段是拍摄与绘画，以此记录场地的表面特征。但这样的记录要求带有设计概念的表达，记录需要显现出对场地空间认知和撷取要素想法。它是一个由表及里的提炼过程，是一种编辑视觉信息和再现复杂系统的练习。这类景观的延伸表现不仅是绘画，更有助于将空间关联到周边环境，或将面对的场地空间关联到新的想法、概念之中。草图、摄影，甚至简单材料都可以综合运用于概念表现，成为意向表达的手段。学生也由此训练抽象式的设计思考能力，发现新的方法去"读"真实的自然景观。

3.追溯"定义"，通过设计互评鼓励运用批评性思维（critical thinking）。

以往的毕业设计课程中学生容易陷入被动的、漫无目的的堆砌设计语言的怪圈。究其原因在于他们并未理清"定义"（Definition），或者说忽略了"定义"的存在。批判性思维和逻辑表达往往是美院学生的短板，但这种能力对于设计师而言却十分重要。

为了补足这一核心教学点，在新的毕业设计课程中我们采用两种教学模式：第一，贯穿于毕业设计课堂教学讨论中的是一种有趣而古老的所谓"苏格拉底诘问法"（Socratic Method）。这种方法来源自古希腊苏格拉底，今天仍广泛运用于美国法学院的课堂教学。其本质就是老师通过谈话方式不断地诘问学生，你对这个问题是怎么看的？那么在这样的情况下你会怎么做？由此达到启发、引导学生的目的，从而形成思辨思维。而通过对话的互相碰撞琢磨砥砺，学生们会对一个问题有更深层次的认识。定义是对一件事物的概念认知与判断，定义是边际也是导向，任何逻辑的起点实际就是"定义"。如果当学生在设计之初就不断追问什么是"建筑遗产"？什么是"合适"？什么是"地缘"？接下来他们就能够用逻辑推导的方法、剥洋葱似的将设计结果逐层呈现。当学生掌握了这种哲学思考方法，他们就能有效解决问题，甚至能产生不一样的思考角度。当然要增强思辨性，还要求学生不断扩大认知边界，扩充历史、人文、社会学、地理学方面的知识，以形成对定义尽量完整的判断。第二，以往的毕业设计老师在指导方案时普遍倾向于"sage on the stage"，即讲台上老师教会学生所有知识。新的毕业设计教学则有意识地培养学生的批判性思维，采取四种教学组织形式：场地现场教学（Field work）、专业教室实际工作（Studio work）、一对一桌前辅导（Desk critique）和每周一次的集体贴图讨论（Pinup）。所谓集中评图机制就是让学生走上讲台做演讲（Presentation），介绍方案概念，同时由其他同学互评方案。改变了单一的老师评价学生模式，由此保证学生积极思考、梳理思路的设计习惯，提升设计评价能力，把毕业设计置于开放的平台中检验"教"与"学"，形成积极互动的讨论风气。这也是国外设计事务所和高校通用的设计评图方式。在美国Wabash College政治学教授Melissa Butler看来，所谓批判性思维就是"通过多角度来检验事物，并积极寻找证据来支持或反驳所读到的观点。不轻信任何说法，通过搜集多角度的资料来检验它们的真实性。"设计与其他通识学科一样，都需要学生通过积极参与和讨论的方式，以积极的姿态自主思考，激发他们尝试以新的方式有效解决问题的勇气。

4.模型教学贯穿始终，作为推导空间的主要手段。

模型在设计过程中起到举足轻重的作用。我院建设有全国一流重点实验教学基地，含有木工实验室、3D模型精雕实验室等，学生可以运用激光雕刻设备制作建筑模型。项目开展之初我们就要求学生制作1:100的主体工业建筑遗存模型。在此基础上学生再运用高密度模型泡沫进行体块切割，探讨室内外空间形态。不论是形体简单的概念模型还是最终的成稿模型，都能为学生提供直观的视觉判断，寻求设计方案的合理性。

五、启示

景观建筑学的毕业设计最为重要的教学目标在于教会学生综合运用几年学习所得的专业知识，行之有效地解决社会当下存在的实际问题。景观建筑学作为服务于社会的应用型综合学科，应当紧扣社会现状展开设计研究，真正承担起景观设计师对于改善生活环境、提升大众生活品质的社会担当。一切游离于社会现状、缺乏对民生问题反思的设计行为构筑的只是"空中楼阁"。让学生直面真实的社会需要，依照规范流程，展开社会性思考、前瞻性规划，将动脑与动手紧密结合才是学科的要求、教育的要求、社会的要求。

"人的行为遵循于他的根源。"（Human behavior flows from the man sources）[（The Story of Philosoph: The Lives and Opinions of the World's Greatest Philosophers y, will Durant.）如果

我们尝试以柏拉图理论的角度追根溯源景观建筑学的设计行为，其灵魂学说的三个核心基石：欲望(desire)、情感(emotion)和知识(knowledge)，很清楚地明示我们如何才能完成称之为最优的"灵魂设计"。以杭州重型机械厂景观更新设计为例，城市CBD综合体新功能注入建筑遗产，满足使用者基本空间下餐饮、购物、博览等各种行为需求是最基本的"欲望层面"的构建；让设计与人产生情感的联系，调动、唤起人们对于重型机械厂的特殊情感则上了一个层级；倘若与此同时还能够讲述一段杭州近现代重型工业发展史及工厂发展过程中曾经发生的人与故事，传递丰富的通史知识，才能称之为用灵魂来设计，这也是设计的最优状态。这与马斯洛（Maslow）"人的需求金字塔模型"中的分层理论异曲同工。金字塔顶层出现的自我实现（"self-realization"）即是作为一个景观设计师自我价值的最终追求。最后所谓的知识（knowledge）本质上就是我们设计教学的核心知识点，即设计行为背后的原理和不断拓展的人文通史知识边界。

学生站在真实的工业建筑遗存前时，面对的已经不单单是一栋栋被弃用的旧建筑，而是一段悠长的、机器轰鸣的工业发展史。他们要做的是思辨地判断该留下什么和该引入什么。学生在本土经济、社会、文化、历史的土壤之上探寻富有独特地域特色的后工业景观更新策略，研究为传达历史信息的工业之后的景观设计语言，为浙江工业废弃地的改造提供新方法，带来新的景观形式。拓展了学生对于景观设计的适应层面、潜在尺度以及社会责任的理解，有利于学生领悟设计背后传承文化脉络的重要意义。这与中国美术学院所倡导的"品学通、艺理通、古今通、中外通"四通人才培养境遇不谋而合。

虽然以此为目标的新教学会遭遇不小的挑战，学生并不习惯于"深挖"而更善于"横向拓宽"。但相信改革后的毕业设计能够给学生提供一个深入反思景观设计意义，发展学生个人对景观的见解和思考的机会。通过艰苦耗时的思考意识到历史遗产背后是文化语境和庞大的文化群体，它是文化定位（Identity）的载体。

六、结语

从2010年到2016年，该课程目前已经摸索了4届。在国际化、数字化时代的今天，新技术赋予我们对于未来无限畅想以及与世界同步的可能性，同时也带来了失去身份定位与文化传统的困惑。城市失去了身份定位与文化形象，变得千篇一律。这时，寻找"身份定位"（Identity）的意义变得尤为重要。正如德国建筑遗产保护委员会副主席鲁道夫·吕克曼教授在其"Can we protect the traditional culture of our life style?"主题演讲中所谈及的两个观点：第一，通过对建筑遗产的研究，我们反思今天的建筑遗产背后人们真实的生活究竟是怎样的？第二，由对建筑遗产的追溯引发我们进一步对于身份认同问题的思考。（"Do we need identity？"）被纳入建筑遗产范畴的工业建筑遗产恰恰是带有一个城市工业发展历史时期烙印的文化象征。笔者试图带领一批批学生借由这类历史遗存载体，去重新找回一个时代和一个城市的一段历史。以景观的手段探寻设计的地缘特性、寻找本土文化的身份特征，可以帮助我们预防文化断层。相信在此层面上的景观设计思考才能够对杭州乃至全国的景观设计教学有所裨益。另外，本课程的教学思路很大程度上受到了德国安哈特应用技术大学建筑系建筑遗产保护专业教学思想的影响，我们希望以杭州工业遗存景观更新为例的景观教学改革能够成为一次独特且具有探索性质的教学尝试，并以此为契机将德国建筑保护教学领域成熟的建筑保护方法与思路植根于杭州本土的文化土壤，寻找到本土身份与文化特征，引发对景观教学的思考。

程银
Cheng Yin

2002—2006年　中国美术学院（环境艺术系）学习，获本科学位。
2006—2009年　中国美术学院中德研究生院攻读建筑设计专业，获柏林艺术大学硕士学位。
2009至今　　　工作于浙江树人大学，并从事建筑设计与室内设计工作。

2002—2006　Architect and interior designer, graduated from the The China Academy of Art Department of environment art.
2006—2009　InUniversity of the Arts Berlin, master's degree at architectural design
Since 2009　Engaged in the architectural design and interior design work after graduation, while teaching at Zhejiang Shuren University Art Institute, engaged in architectural design teaching so far.

解析德国威廉皇帝纪念教堂的场所体验
——用现象学的方法

程银

【摘要】本文通过现象学的方法研究"二战"重建后的威廉皇帝纪念教堂建筑群,用现象学中的场所体验来分析教堂建筑的整体处境、内部空间关系以及建筑的细节,以此来阐述新旧建筑的关系,探讨建筑遗产保护的方法,以期对建筑遗产保护产生积极的影响。

[Abstract] In this article, we research the building group of Kaiser Wilhelm Gedaechtniskirche rebuilt after the Second World War with phenomenology and analyze the whole condition, internal spatial relationship and the details of buildings with the site experience in phenomenology in order to explain the relationship between new buildings and old buildings and discuss the methods for architectural heritage protection and hope to produce some positive influence on architectural heritage protection.

一、引论

近年来,哲学理论对当代建筑理论产生了深远的影响,其中最重要的是现象学、当代美学、人类学等。在各种建筑理论家与哲学家的讨论中,经常将现象学的方法运用于建筑空间的研究之中。这里所讨论的"现象学"是德国哲学家爱德蒙德·胡塞尔(Edmund Husserl,1859—1938)在20世纪10年代开创的哲学运动。胡塞尔以后出现过很多现象学哲学家,他们的现象学哲学理论都建立在胡塞尔的研究方法之上。其中,对建筑理论影响较大的现象学家是德国哲学家马丁·海德格尔(Martin Heidegger,1889—1976)和法国哲学家莫里斯·梅洛-庞蒂(Maurice Merleau-Ponty,1908—1953)。他们的现象学方法对以后多位建筑理论家以及建筑师产生影响,其中包括:诺伯格-舒尔茨(Christian Norberg-Schulz)、帕拉斯马(Juhani Pallasmaa)、斯蒂文·霍尔(Steven Holl)等。

本文将对哲学现象学中可用于建筑空间研究的场所体验做一概述,并尝试以现象学的方法来解析二战重建后的德国柏林威廉皇帝纪念教堂建筑群中的空间体验,通过分析旧教堂遗址和新建建筑组群的空间关系及人在建筑空间中的场所体验来学习建筑遗产保护的方法,以期对建筑遗产的保护与修复以及传统空间的现代表达产生积极影响。

二、现象学中场所体验的建筑理论

现象学(phenomenology)原词来自希腊文,意为研究客观事物的外观、表面迹象或现象的科学。即探究事物的本质,现象"还原"。建筑现象学大致有两个研究方向:一是侧重于纯学术理论研究领域的以胡塞尔和海德格尔为源头的存在主义现象学。另一个是侧重于建筑设计和实践的以莫里斯·梅洛-庞蒂为代表的知觉现象学。

在海德格尔的现象学中,场所精神是天、地、神、人的集中体现。无论是人为的场所,还是自然的场所,一旦存在于世界,与人构成某种关系,就必然存在这种场所。场所与建筑和城市空间密切相关,场所精神存在于能够容纳体验,能够产生共鸣的空间之中,场所精神是空间体验的产物,也是空间的再创造,这种再创造取决于人,即不同人的生活经验产生不同的场所体验,这也是建筑现象学的核心精神。

诺伯格在胡塞尔与海德格尔的基础上,对现象学进行了建筑化和图象像化的解释。他在《场所精神》中强调了"场所"(place)在建筑设计中的重要性。只有当人们体验了场所和环境之后,他才"定居"

（dwell in）了，"居"意味着生活发生的空间，这就是"场所"。而建筑的存在目的就是使得原本抽象，无特征的同一均质的"场址"（site）变成真实、具体的人类行为发生的"场所"。他的现象学的研究方法是：立足于真实可靠的"生活世界"中，更加关心人的行为和体验，通过对生活的观察和环境的体验，使得建筑更具人性化。[1]

莫里斯·梅洛-庞蒂在他的《感知现象学》中指出：现象学既是对本质的描述，又是对本质的进一步实现。这种实现是永无止境的，可以说，现象学是对体验的深度和广度的不断发掘。人应该通过声响、寂静、气味、触摸的形状以及肌肉和骨骼的知觉来感受，使用者处于建筑中并对建筑有更深层次的理解、深化，通过亲身体验有自己的二次创造。"知觉"，正是人们通过自身的体验，在先于科学解释之前，不带有任何哲学偏见地进入"感知世界"的途径。[2]

三、用现象学的方法解析二战重建后的德国威廉皇帝纪念教堂的场所体验

分析一个建筑，最理想的方法是现场考察，以获得第一手的直观资料；次之是访谈，以获得设计师的设计思路；再次之是找到确凿的记录教堂历史的文献。2008年笔者在德国柏林艺术大学学习期间，每天都要路过威廉皇帝纪念教堂，对教堂周围环境以及内部空间有过较为深刻的体验，并且收集了教堂内大量文献资料。因此，笔者用的是第一种与第三种方法。

图1（左）"二战"前的威廉皇帝纪念教堂　　　　图2（右）"二战"后被炸毁的教堂塔楼遗址

威廉皇帝纪念教堂（Kaiser-Wilhelm-Gedaechtnis-Kirche）是19世纪末德意志帝国皇帝威廉二世为纪念他的祖父、德意志帝国的第一个皇帝威廉一世而建，因而被命名为"威廉皇帝纪念教堂"（图1）。教堂位于德国柏林市繁华地段布赖特沙伊德（Breitscheidplatz）广场上，西面是Zoo火车站，于1895年9月建成，是一座带有哥特元素的新罗马式风格的建筑，教堂的尖顶的最高点是113米，是柏林当时最高的建筑。著名的艺术家给这座教堂建筑上创作了马赛克、浮雕以及雕塑使该建筑成为当时西柏林的一个亮点以及地标性建筑。1943年11月23日，教堂在二战盟军轰炸中严重损毁，原来113米高的尖顶倒塌，只剩68米(图2）。战后，在教堂的修建工作中，68米高的旧教堂钟楼残骸被保留和保护下来，因为德国人认为，断裂的拱顶与教堂周围的遗址，就是一座反对战争的纪念碑。教堂的修建工作出自卡尔斯鲁厄大学建筑学教授埃贡·艾尔曼（Egon Eiermann，1904—1970）之手，他计划将残骸完全拆除并建造一座现代风格的新教堂。然而这一设计引起了柏林市民的极大争议和反对，柏林市民希望保留旧教堂的残骸以示纪念。最终双方达成了妥协，68米高的旧教堂钟楼残骸得以保留和保护，以警世战争，残骸周围则依照艾尔曼的方案建造四栋新建筑：八边形的教堂中殿、六边形的钟楼、四边形的礼拜堂以及前厅（图3）。艾尔曼也为教堂内室设计了圣坛、洗礼池和蜡烛台等。

1.人对教堂建筑整体处境的体验

斯蒂芬·霍尔（Steven Holl)在其著作《锚固》（Anchoring）中提出："建筑与处境有密切关系"[3]，并讨论到如何将建筑坚实地植根和锚固于建筑独特的场所中 。[4]建筑必须与周围的环境、历史、文化以及使用者发生联系，一旦建筑脱离了场所，移植到其他地方就会隔断这种联系，隔断了场所精神，人的场所体验也就截然不同。二战以前，威廉皇帝纪念教堂的历史是一个相对单纯的为纪念威廉皇帝而建造的，承担着教堂的功能。随着时间的流逝，教堂的时间印迹逐渐厚重，二战中被损毁的那一刻，教堂已经不再是单纯意义上的教堂，它那断裂的尖顶被瞬间锚固在这个场所，并和当时被战争破坏的城市、惊恐的市民一起被锚固在那个时间：1943年11月23日晚。教堂所承载的场所体验瞬间发生了历史性的变化，因为建筑的场所体验是历史文化的记忆与现在和未来的一种无限的联系。破损的教堂不仅告诉全世界德国人对待战争的态度，更是一座反对强权、反对战争的纪念碑，它存在的精神意义早已大于实际使用功能，它已经融进了历史中。如果当初柏林市民选择拆除旧教堂，完全建造一座新教堂，建筑场所中对历史文化的记忆就会发生变化，那么人的这个空间中的场所体验就会完全不同，人体验不到二战中曾经发生过的历史，后代德国人对待战争的态度也会发生变化，整个民族精神也会截然不同。所以它必须存在，哪怕它已失去了空间的物理使用功能。

根据精神分析理论家西格蒙德·弗洛伊德（Sigmund Freud,1856—1939）的理论，体验是一种瞬间的幻想，是对过去的回忆，既是对过去曾

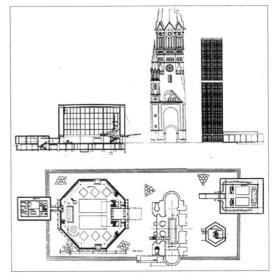

图3 新建教堂组群平面图和剖面图

图4（左）教堂新旧建筑之间的关系，图5（右）八角形建筑外墙和内墙的钢结构链接

经实现的东西的追忆，也是对现在的感受，是早年储存下来的意象的显现；同时，又是对未来的期待，以回忆为原型瞻望未来，创造美景或幻想。因此，体验是一种将过去、现在和未来联系起来，并综合为建筑空间的再创造。[5] 建筑也与它所处的地理与社会环境有关，只有既体验产生和形成建筑的地理和社会环境，同时也体验建筑，才是完整的建筑空间。1961年，新教堂落成，教堂新建部分由4幢新建筑组成：八边形的教堂中殿、六边形的钟楼、四边形的礼拜堂以及前厅。老教堂的塔楼废墟在建筑群的中央，使新建筑和老建筑之间构成充满张力的统一（图4）。这样的新旧对比在柏林这个城市中，与它的社会环境一起构成一幅过去、现在、未来的图像。新的建筑带来的是新的时代、新的技术，完全区别于老建筑。新的建筑群将过去、现在与未来的空间体验联系起来，人的过去回忆与当下体验同时显现，形成人对建筑空间的新的体验，即现象学概念中的再创造。

2、人在教堂内部空间中的体验

八边形的教堂中殿平面呈现八角形，双层墙体，内墙与外墙用钢架链接，两片墙体由相同的方法构造。墙体结构由钢柱支撑，两片墙中间是2.45米的回廊，起很好的隔音作用，完全隔音的双层墙体使建筑处于闹市区也可以获得非常安静的室内祈祷空间（图5）。整个建筑的高度为20.5米，直径35米，放置可以容纳1000多人的位置。钢结构的建筑体立面上的水平线和垂直线通过钢带连接，在这些钢材料的格子状结构中被插入几乎正方形的水泥蜂房，水泥蜂房的小格子里镶嵌着法国玻璃艺术家加布里尔·洛伊尔（Gabriel Loire）制作的彩色玻璃（图6）。这个建筑群因为马赛克彩色玻璃而出名。这种玻璃在当时是一种全新的玻璃艺术工艺。整幢建筑一共使用了16000块马赛克彩色玻璃。为了这个教堂，加布里尔特别研究出一种厚实而带色彩的玻璃的工艺，他将蓝色调进行变换与组合，使每块玻璃上的颜色组合都是独一无二的，"蓝色是和平，红色是友谊"，他这样解释玻璃上的色彩，这是对当时特殊的社会环境下的体验。这些色调在

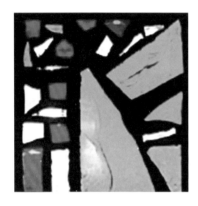 

图6（左）马赛克彩色玻璃窗，图7（右）新建教堂室内

太阳光的作用下折射出非常特别的彩色光线。玻璃在烧制出来后就被镶嵌在水泥蜂房中做成正方形的水泥板制成品，再被运回柏林安装在钢架上。在建筑设计中，这种彩色玻璃窗是传统玫瑰花窗的现代表现形式（图7）。白天，阳光通过彩色玻璃进入教堂内部，形成斑驳的变化丰富的蓝色光线，这种彩色玻璃窗像哥特式教堂里的玫瑰花窗那样将教堂应有的庄严，宗教崇高的精神表达出来。夜晚，教堂内的光线被彩色玻璃折射出去，整个教堂的外立面呈现出斑斓的蓝紫色（图8）。彩色玻璃的原型是玫瑰花窗，人的过去记忆与当下的体验结合起来，形成一种全新的教堂空间的体验，人的思维同时穿过彩色玻璃窗和过去的岁月，将身体与知觉结合了空间与时间，形成一种新的场所体验。人在建筑内部的空间体验就如同站在雄伟高大的哥特式教堂中那样，完全能体会到教堂建筑空间的传统精神。新的教堂建筑的内部空间所传达的精神气质不变，但是传递精神的载体变了，即建筑的材料、表现方式是现代的，空间精神是传统的，这就是传统建筑空间在新建筑空间中的完美演绎。

新建筑组群中的设计统一而富有变化，在四边形的礼拜堂的设计中，外墙同样是钢架结构，被镶嵌玻璃的水泥蜂房填充，但是内墙与外墙之间留出一圈院落空间，全玻璃的内墙可以完全感受到两片墙之间的院落。这种区别而又有联系的空间处理方式使人的空间体验保持了一种完整性。

3.人对建筑空间中细节的体验

被保留下来的旧教堂是一个正在使用的、让人参与其中的建筑。旧教堂的大厅里展示了威廉一世丰富而又多变的生活经历，由砂岩材料制作的马赛克拼贴头像和用大理石雕刻的浮雕仍然可以清晰地传达出人像的信息（图9）。大厅北面用16块展示牌记录了1943年11月23日晚上以前的所有历史。老教堂祭坛上的耶稣基督像旁边的"考文垂的十字架"（Nagelkreuz von Coventry）和德国艺术家库尔特·罗伊博（Kurt Reuber，1906—1944）的画作"斯大林格勒的麦当娜"（Madonna von Stalingrad）是反对战争的证明。这两件物品分别来自英国和俄罗斯的赠送，意味着和解。在老教堂的遗址里，每周五下午一点，信徒们用考文垂（Coventry）大教堂的安抚经代替午时经，也表示和解。建筑遗址加上历史事件以及与历史相吻合的宗教活动，德国对待战争的态度会被全世界所接纳，跟建筑的场所精神与人在此空间中的体验不无关系。这样的空间布置贯穿着历史文化，是人产生正确的场所体验的基础，倘若老教堂遗址被拆除，场所精神就会被割断，人们还会有这样贯穿历史直面战争的场所体验吗？

图8（左）教堂夜景，图9（右）砂岩材料拼贴的威廉皇帝像

优秀的建筑空间中，人应该既是旁观者又是参与者，人会接收建筑空间氛围的感染。芬兰建筑师及现象学家尤哈尼·帕拉斯玛(Juhani Pallasmaa)提出了对建筑的精神上的理解。他认为"要开拓视野，看到由感觉、梦想、忘却的记忆以及想象所组成的第二个现实世界。"这种第二个现实世界，就是体验的世界，只有体验才能产生场所精神。[6]

四、结束语

古建筑是历史的载体，真实地呈现建筑的场所、空间、结构、细节，是对人类历史的尊重。"二战"结束后，德国有大批的古建筑被损毁，类似威廉皇帝纪念教堂的建筑保护方法在其他很多建筑中也使用过。通过以上对威廉皇帝纪念教堂的分析，建筑遗产保护的方法，从现象学的角度归结起来有以下几方面：首先，一个受保护的古建筑应该是被使用的建筑，而不是像是被陈列在博物馆里让人观赏的建筑。人是建筑空间的参与者，只有人的参与与体验，建筑才会产生场所精神。其次，新加建的建筑应该是新的工艺、新的表现手法，新旧部分是非常容易被识别的，不能为了取得与老建筑的表面的和谐而去新建一个那个时代的建筑，那充其量只是一个赝品，模糊了历史的真实性，混淆了人在场所中对历史、文化以及时间的正确体验。再次，尊重建筑遗产周边的环境，要将这一区域作为一个整体来保护。建筑空间的场所体验是一种将过去、现在和未来联系起来，与所处的地理与社会环境联系起来，再综合为建筑空间的再创造的过程。

注释

[1] http://www.gn-art.com/zixun/archives/73986
[2] http://www.gn-art.com/zixun/archives/73986
[3] Holl S.Anchoring[M].New York:Princeton Architectural Press, 1989, P. 9.
[4] 沈克宁，建筑现象学[M]. 北京：中国建筑工业出版社，2008, P. 168.
[5] 彭怒 支文军 戴春，现象学与建筑的对话[M]. 上海：同济大学出版社，2009.7，P. 271.
[6] 彭怒 支文军 戴春，现象学与建筑的对话[M]. 上海：同济大学出版社，2009.7，P. 272.

主　　编：杨修憬　　　　　　　　　Chief Editor: Yang Xiujing
责任编辑：徐新红　　　　　　　　　Resporsitte Editor: Xu Xinhong
执行编辑：戚晨曦　项亚量　　　　　Executive editor: Qi Chenxi, Xiang Yaliang
整体设计：汪建军　王惠宁　　　　　Book Design: Wang Jianjun, Wang Huining
翻　　译：陈　珊　金　鑫　　　　　Transtation: Chen Shan, Jin Xin
责任校对：朱　奇　　　　　　　　　Resporsible Proofreader: Zhu Qi
责任印制：毛　翠　　　　　　　　　Resporsible Publisher: Mao Cui

图书在版编目（CIP）数据

中国美术学院建筑遗产保护国际论坛论文集 / 杨修憬主编. -- 杭州：中国美术学院出版社，2017.1
　ISBN 978-7-5503-1260-9

Ⅰ．①中… Ⅱ．①杨… Ⅲ．①建筑－文化遗产－保护－文集 Ⅳ．①TU-87

中国版本图书馆CIP数据核字(2017)第017013号

中国美术学院建筑遗产保护国际论坛论文集
杨修憬　主编

出 品 人：祝平凡
出版发行：中国美术学院出版社
地　　址：中国·杭州市南山路218号 ／ 邮政编码：310002
网　　址：http://www.caapress.com
经　　销：全国新华书店
制　　版：杭州海洋电脑制版印刷有限公司
印　　刷：浙江省邮电印刷股份有限公司
版　　次：2017年7月第1版
印　　次：2017年7月第1次印刷
印　　张：11.75
开　　本：787mm×1092mm　1／16
字　　数：230千
图　　数：450幅
印　　数：0001－1100
书　　号：ISBN　978-7-5503-1260-9
定　　价：78.00元